Mamadou Barry

Coastal Erosion and Impacts

Mamadou Barry

Coastal Erosion and Impacts

in the Commune of Kafountine Basse Casamance

ScienciaScripts

Imprint
Any brand names and product names mentioned in this book are subject to trademark, brand or patent protection and are trademarks or registered trademarks of their respective holders. The use of brand names, product names, common names, trade names, product descriptions etc. even without a particular marking in this work is in no way to be construed to mean that such names may be regarded as unrestricted in respect of trademark and brand protection legislation and could thus be used by anyone.

Cover image: www.ingimage.com

This book is a translation from the original published under ISBN 978-613-8-43074-2.

Publisher:
Sciencia Scripts
is a trademark of
Dodo Books Indian Ocean Ltd. and OmniScriptum S.R.L publishing group

120 High Road, East Finchley, London, N2 9ED, United Kingdom
Str. Armeneasca 28/1, office 1, Chisinau MD-2012, Republic of Moldova, Europe
Printed at: see last page
ISBN: 978-620-6-10145-1

Table of contents :

ASSANE SECK UNIVERSITY OF ZIGUINCHOR

UFR SCIENCES AND
TECHNOLOGIES DEPARTMENT
GEOGRAPHY

COASTAL EROSION AND IMPACTS IN THE COMMUNE OF KAFOUNTINE (LOWER CASAMANCE)

By:

Mamadou BARRY

DEDICATION

Praise be to Almighty Allah for allowing me to carry out this work.

I dedicate this memoir :

To my late uncle Ismaila Barry, may the earth be light to him;

To my mother Fatoumata Barry and my father Ibrahima Barry ;

To my uncle Famara Mané;

To my brothers, sisters, cousins;

ACKNOWLEDGEMENTS

It is with pleasure and gratitude that we take this opportunity to express our gratitude to all those who have contributed to the success of this work.

I would like to thank :

- to Pr Amadou Tahirou Diaw for agreeing to supervise this work;
- to Dr. El Hadji Balla Dièye, for agreeing to supervise this work. His great availability, his constant support, his very constructive criticism and his sound advice were invaluable to me throughout these years of research. May peace be with him;
- to the members of the jury for agreeing to judge this research work,
- to the entire teaching staff of the Geography Department of the Assane Seck University of Ziguinchor for the quality of their teaching and their availability. I would like to mention Dr. Tidiane Sané, Dr. Ibrahima Mbaye, Dr. Omar Sall, Dr. Alvares G. F. Benga, Pr. F. Benga, Pr Oumar Sy, Pr Pascal Sagna, Dr Aidara Ch. A.L. Fall, Dr Cheikh Faye, Dr Abdourahmane M. Sène ; Dr Alla Manga ;
- the agents of the Abéné Marine Protected Area, Commandant Mamadou Sidibé, Lieutenant Mamadou Faye, Seckou Moussa Sagna and Yaya Faye for their support and guidance;
- to all the people from whom I was able to gather the information that enabled me to carry out this work. I would like to thank my uncle Abdoulaye BARRY for welcoming me and putting me up during my stay, Mamadou Yacine Ba for accompanying me in the field, and the village chiefs of Hilol Dominique Demba, Boko Monsieur Diassy and Olpa Diassy of Saloulou;
- to my grandmother Fatoumata Sow, my cousins Mamadou A. and Cheikh A.T. Mané, and my friends Mohamet Ndiaye, Mamadou Diely Diallo, Abasse Bibi, Ousseynou and Assane Diédhiou from Ziguinchor for all their support;
- the doctoral students of the UASZ Geography Department, in particular Mamadou Thior and Boubacar Solly, for their advice and encouragement;
- to my classmates from Assane Seck University, Seny Diatta, Balla Dia, Eric S. Kamony, Bissanagha A. Dos Santos, Seydou and Alassane Baldé, Moustapha M. Diédhiou, Moussa Sané, Alexandre Badiane, Jean Baptiste Biaye, Kader Baldé, Ana Gomis.

SUMMARY

The coastal zone of the Commune of Kafountine is at risk of marine erosion as a result of climate change combined with anthropogenic pressure.

The aim of this study is to analyze the factors driving coastal erosion in the Commune and its impacts. The methodological approach is based on a diachronic analysis of coastline evolution using a series of four Landsat satellite images (Landsat MSS 1979, Landsat TM 1986, Landsat ETM+ 2000 and

Landsat OLI_TIRS 2015).
Between 1979 and 1986, the results showed a general trend towards progradation. On the other hand, during the periods 1986-2000 and 2000-2015, the coastline showed contrasting trends, marked by retreating and advancing phases. This dynamic is the result of coastal development and climate change.
The resulting environmental and socio-economic impacts are varied (loss of rice-growing land, destruction of tourist and social infrastructures, degradation of water quality and vegetation). This situation has prompted the local population to adopt adaptation strategies in the face of coastal erosion, with the support of state structures (AMPA and Eaux et Forêts) and NGOs (Justice et Développement and AZOHS). These adaptation strategies are based on mangrove reforestation and coastal dune fixation. They are deemed insufficient in a context of permanent change in the coastal zone of Kafountine Commune. It is obvious to suggest a few recommendations with a view to sustainable coastal management.

Key words: Coastal erosion, Impact, Littoral, Strategy, Kafountine, Basse Casamance, Senegal, Islands.

ABSTRACT

The coastal zone of the township of Kafountine is subject to the risk of marine erosion due to climate change combined with anthropogenic pressure.

The aim of this study is to analyze the factors of coastal erosion of the Commune and its impacts. The methodological approach is based on a diachronic analysis of the evolution of the coastline from a series of four images of the Landsat satellite (Landsat MSS 1979, Landsat TM 1986, Landsat ETM + 2000 and Landsat OLI_TIRS 2015).

The results showed a general tendency towards progradation between 1979 and 1986. On the other hand, the periods 1986-2000 and 2000-2015, the coastal region experienced a contrasting evolution marked by backward phrases and sentences of progress. This is a consequence of coastal development and climate change.

The resulting environmental and socio-economic impacts are of various kinds (loss of rice land, destruction of tourism and social infrastructure, degradation of water and vegetation quality).

This situation led to the population to undertake strategies of adaptation for coastal erosion with the support of state structures (AMPA and Waters and Forests) and of the NGO (Justice and of development and AZOHS). These strategies of adaptation are based on the reforestation of the mangrove swamp and the fixing of the coastal dunes. They are considered insufficient in a context of permanent changes in the coastal zone of the township of Kafountine. It is obvious to suggest some recommendations from a perspective of sustainable coastal management.

Key words: Coastal erosion, Impact, Littoral, Strategy, Kafountine, Casamance, Senegal.

GENERAL INTRODUCTION

The coasts, where land, sea and air meet, are narrow bangs characterized by high mechanical energy and great biological diversity, attracting more and more human covetousness every day. Unfortunately, most people are unaware of the true nature of these coastal areas. Multi-faceted and sometimes conflicting activities develop there, often destabilizing these precarious environments (Savadogo, 2013).

Morphological and sedimentary imbalances induced by ill-considered and ill-advised development projects exist all over the world. In most cases, costly and not always effective solutions are used to try and remedy the damage, which can be irreversible and which could undoubtedly have been avoided with a better understanding of the coastal environment.[1]

In Senegal, the 706.72 km of coastline are no exception (Diaw, 1997). This coastal strip constitutes a privileged geographical space where socio-economic activities are highly attractive and where a rich, varied and vulnerable environmental heritage is to be found (Ndao, 2012). However, this highly attractive coastal area, with its concentration of human activity, has led to an imbalance, accelerated coastal erosion and negative impacts on the environment and the economic life of local populations.

The factors responsible for coastal erosion are varied and often debatable. Most authors (Diallo, 1982; Sall, 1982; Soumaré, 1992; Faye, 1993; Niang-Diop, 1995; Diaw, 1997; Diara, 1999; Adjoussi, 2000; Seck, 2001; Sy, 2013) maintain that climate change, sea-level rise, storms and variations in natural sediment supply are the main causes of coastal erosion. Anthropization of the coastal environment contributes to the acceleration of this phenomenon.

The resulting environmental consequences can be seen in the general degradation of coastal and marine ecosystems, leading to retreating coastlines, loss of biodiversity and rising salt water levels. These environmental consequences can influence socio-economic activities. They can lead to the displacement of populations from one locality to another, and to losses of

land and property, particularly in regions where economic activity makes a significant contribution to local income.

Understanding the problems affecting the environment and socio-economic activities in the Commune of Kafountine is therefore essential to the study of coastal erosion.

This memoir is divided into three main parts:

- a first part dealing with the theoretical framework and research methodology;

[1] Guide to adaptation options in coastal zones for local decision-makers, UNESCO/IOC 2012

- a second part devoted to +characteristics of the coastal zone and erosion factors ;
- a third section that assesses the impacts of coastal erosion and analyzes adaptation strategies.

PART ONE :

THEORICAL FRAMEWORK AND METHODOLOGY SEARCH

A theory is a set of propositions used to logically unify concepts in order to explain and interpret certain aspects of the reality we seek to account for. In science, the aim of a theory is to understand and explain real-world phenomena. In this section, we present the theoretical framework on which our research is based, and the methodology applied to arrive at the results.

CHAPTER 1: THEORETICAL BACKGROUND

Before reviewing the literature on our subject, we need to explain the reasons behind our choice.

I. Issues

Coastal erosion is a well-known phenomenon worldwide (Bird, 1985; Paskoff, 2001). In the context of climate change, this phenomenon is set to increase, notably due to the global rise in sea level and the increased intensity of storms (IPCC, 2001). Indeed, more than two-thirds of the world's sandy coasts are thought to be eroding (Nichols, 1998), and coastline retreat currently affects more than 70% of the world's beaches (Bird, 1985). According to McCarthy *et al.* 2001, 40% of the world's population live within 100 km of the coastline, and this phenomenon is likely to have a major impact on them.

In Europe, erosion is estimated to affect 40% of beaches (European Union, 2004) and over 50% of sandy coasts in mainland France (IFEN, 2006). In the United States, at least 66% of the sandy coastline of the Gulf of Mexico is said to be in retreat (Morton *et al.*, 2004), 45% of beaches from Florida to North Carolina (Morton *et al.*, 2005) and 40% of Californian beaches (Hapke *et al.*, 2006).

In Latin America, 81% to 84% of beaches in the state of Rio Grande do Sul (Brazil) are said to be eroding (Esteves *et al.*, 2002 *in* Dillenburg *et al.*, 2004). Erosion phenomena have also been demonstrated on Indian beaches in the Bay of Bengal (Gopinath and Seralathan, 2005). Pirazzoli (1993) estimates that, at the current rate of progress, small island states in the Indian Ocean (such as the Maldives) could be wiped off the map. On the African continent, coastal erosion is a worrying problem. On average, 1 to 2 m of beach per year is swallowed up by the Atlantic Ocean in the coastal countries of West and Central Africa (UNEP/UNESCO/UN-DESA, 1985, Ibe and Quelennec, 1989).

In Senegal, the problem of coastal erosion is worrying, with recession rates varying between 1 and 2m/year (Niang-Diop, 1995). This is the case with the opening of the Lagoba breach in the Sangomar spit, which caused recession rates of 198 to 264 m/year between 1987 and 1991 (Diaw, 1997). This breach reached an advance rate of 164 m/year in 2010 (Dièye *et al.*, 2013). In Dakar, the

Yoff coastline has experienced an average retreat rate of around 1 m/year since 1945, while in the Diamalaye sector, the coast is eroding at a rate of over 2 m/year (Sy, 2013). In the Rufisque area, Diallo (1982) and Sall (1982) determined an average recession of around 1.30 m/year for the period from 1933 to 1980. On the Langue de Barbarie, the market gardening areas once farmed by the inhabitants of Dounbaba Diéye Island are disappearing as a result of the ever-increasing width of the Langue de Barbarie breach, which now exceeds 1,000 m, causing the collapse of 14 houses in Goxxumbacc (Sy *et al.,* 2010). On the agricultural front, market gardening in the Niayes of Gandiolais has been compromised by rising salinity, accentuated by pressure on the water table. This situation has led to falling yields and even the abandonment of many adjacent gardens in the cordon vif (Diatta, 2004).

In Casamance, the phenomenon of coastal erosion seems to have gained momentum in recent decades. Socio-economic impacts include damage to the built environment, disruption to economic activities and higher maintenance costs for coastal infrastructures (Diatta, 2012; Sarr, 2012). Ecologically, they also manifest themselves in the loss or modification of habitats, the effects of which will be felt on the traditional activities of coastal communities (Diatta, 2012; Thior, 2013).

In the Commune of Kafountine, the morphology of coastal forms is evolving under the influence of natural and/or anthropogenic actions. The consequences of rising sea levels are clearly visible on the coast. In particular, they can be seen in the retreat of the shoreline. Between 1979 and 2000, Faye (2010) reported retreats of -13.11 m/year and -24.03 m/year (Ebédou and Kalissaye) and -17.19 m/year and -14.07 m/year between Diogué and Nikine. The consequences can be seen along the coastline, with the collapse of tourist facilities. The islands of the Casamance estuary (Diogué, Boko, Hilol and Niomoune) are experiencing an upsurge in marine erosion, salinization and acidification of arable land, and a recurring problem of access to quality water. These islands cover a highly sensitive physical environment: they are fluvial islands with a substrate composed of sand and tannin on both sides, and a predominantly mangrove ecosystem. In these amphibious areas, salinization is increasing rapidly, making rice cultivation less and less feasible. The little usable space on certain islands is under threat from the advance of the sea, facilitated by the subsidence of the islands but also by anthropogenic factors such as mangrove exploitation.

The socio-economic and ecological importance of the West African coastal zone in general, and of sandy coastlines in particular, in a context of environmental crisis, justifies a reflection on the functioning, evolution and management of the use of coastal resources. With this in mind, the Programme Régional de Conservation des Zones Côtières et Marines en Afrique de l'Ouest (PRCM) was launched in 2001 to support the creation and co-management of Marine Protected Areas. This same initiative was undertaken in 2004 by the State in the Commune of Kafountine. In fact, the creation of the Abéné Marine Protected Area aims to conserve marine and coastal biodiversity and improve socio-economic benefits for local populations. Local authorities, NGOs (Justice et Développement and AZOHS) and the Water and Forestry Service are also working to protect and conserve the coastal and marine environment.

II. State of the art

Coastal erosion is a natural phenomenon that has been occurring on coasts for thousands of years. It affects 70% of the world's sandy beaches (Bird, 1985; Paskoff, 2001; Niang-Diop, 1995) and is the result of a combination of natural and human factors.

A worldwide synthesis study by Bird (1985) showed that the recent trend of beaches and coastlines is exposed to erosion in 70% of cases, while only 10% are advancing. The remaining 20% show relative stability. Since rates are by definition forms of accumulation, their current erosion reflects a reversal of their evolutionary rate.

According to the United Nations Framework Convention on Climate Change (1992), climate change is attributed, directly or indirectly, to human activities that alter the composition of the global atmosphere, adding to the natural variability of the climate. This phenomenon is expected to continue for at least 100 years. For the entire Senegalese coastline and for a sea-level rise of 1m by 2100. Dennis *et al* (1995) predicted that between 55 and 86 km^2 of beaches would disappear and 6,000 km^2 of low-lying areas, mainly estuaries, would be flooded, while on the mainland (land), an extension of arid areas following an overall decrease in rainfall is predicted. Coastal erosion remains a phenomenon that spares no coastal state.

In 2004, some 20,000 km of coastline, representing 20% of the European Union's total coastline, were particularly affected by coastal erosion (European Union, 2004). In France, an estimated 1,800 km of coastline are eroding at an average rate of 0.1 to 1 m (Mate and Metl, 1997).

In Québec maritime, the work of Boudjera (2010) has shown that the risk of coastal erosion is truly serious, and coastal degradation represents one of the greatest environmental problems facing the Côte-Nord region. Coastal erosion particularly affects the St. Lawrence estuary, which stretches from Quebec City to the Iles-de-la-Madeleine. Indeed, in these areas, 70% of the coastline is made up of unconsolidated formations that are highly vulnerable to erosion by wave action, swells and other factors.

In West and Central Africa, the United Nations Environment Programme (UNEP), as part of its Regional Seas Program in 1985, identified the coastal erosion problems facing the 21 countries in the region. This revealed that shoreline recession rates were relatively high in the Gulf of Guinea (1 to 5 m/year) between Côte d'Ivoire and Nigeria. These trends were also reported in the synthesis work carried out as part of the vast regional project on the Gulf of Guinea Large Marine Ecosystem, piloted by UNIDO in 1998. The marine and coastal environment of Cape Verde, Mauritania, Gambia, Guinea-Bissau and Senegal is a highly productive ecosystem with significant marine biodiversity. It supports a significant proportion of the livelihoods of coastal communities. However, assessments set out in national communications to the United Nations Framework Convention on Climate Change (UNFCCC) and in the Second Assessment Report of the Intergovernmental Panel on Climate Change (IPCC, 2007), concluded that "widespread coastal erosion due to climate change is one of the most serious environmental problems facing the region". While it is true that coastal erosion and sedimentation have been a reality in these countries for centuries (ACCC/UNESCO, 2012), both processes are strongly influenced by changing climatic conditions. The resulting changes in hydrographic and oceanic conditions due to climate change are likely to exacerbate coastal erosion and sedimentation problems in the West African region.

Ibe and Antia (1983) reported coastal erosion due to dam construction in the Niger Delta of Nigeria. They also indicated that the growing problem of erosion on the barrier coast and lagoons of western Nigeria was due in part to the construction of agricultural dams on rivers such as the Ogun and Oshun, which previously brought sand to the coast. Ly (1980) also pointed out that the Akosombo dam on the Volta in Ghana is causing or accelerating the general erosion of Ghana from the center to the east, and to the east of the river mouth. According to Epiphane (2009), coastal erosion has serious socio-economic consequences in West and Central Africa. In Côte d'Ivoire, it has led to the relocation of the town of Grand Lahou 15 km across the lagoon, and destroyed colonial and historical monuments and communication routes in Port-Bouet (Savadogo, 2013). In Ghana, particularly in Accra, tourist facilities at Labadi Beach and buildings in Saint James were destroyed (Thiam and Crowley, 2014).

In Senegal, studies on the dynamics of coastal erosion have been carried out mainly on the northern coasts (Sall, 1982; Diop 1990; Niang-Diop, 1995; Diaw, 1997; Adjoussi, 2001; Faye, 2010; Faye, 1993). Their studies have shown that Senegal boasts significant ecological diversity, with plant formations in the Niayes in the north and mangrove vegetation lining the *bolongs in the* Saloum Delta estuary, the Senegal estuary and Casamance. Today, these coastal zones are threatened by coastal erosion. Hydrodynamic agents and strong demographic pressure are the main culprits. Niang-Diop (1995) shows that coastal erosion phenomena can have more than just natural causes. Human activities can interfere with the natural agents controlling the coastal zone. I1 can intervene in two main ways: either by modifying sediment supply through the construction of dams on rivers, sand extraction on beaches or offshore, hardening of the shoreline, destruction of littoral vegetation; or by interfering with littoral transport, essentially through the construction of structures perpendicular to the shoreline (dikes, jetties, groins, breakwaters, etc.) which can interrupt or halt part of the sediment transport effected by littoral drift. This hypothesis is supported by Sy, (2013) who states that the factors influencing sediment processes (sediment transport and coastline movement) are coastal morphology (sediment characteristics, coastline layout) and human factors (occupations and uses).

The consequences can manifest themselves in retreating coastlines and threats to socio-economic infrastructures. For example, Sall (1982), based on a survey of marker posts on the Hydrobase-Tassinére axis, obtained an average recession rate of around -1.03 m/year between 1972 and 1978. The thesis work of Niang-Diop (1995) and Faye (2010) is a fairly accurate source of information. Niang-Diop (1995) used images from 1954, 1975 and 1989, taken during the dry season and on dates that did not coincide with exceptional events. According to this author, in Saint-Louis, the average rate of recession for the period 1954-1989 was -0.50 m/year, with zones of maximum erosion moving from one end to the other. Between 1954 and 1975, the Ndar-Tout sector retreated significantly (-1.20 m/year), while the Guet-Ndar sector remained relatively stable (-0.30 m/year). Between 1975 and 1989, the Ndar-Tout sector remained virtually stable (-0.20 m/year), while the Guet-Ndar sector experienced significant erosion (-3.20 m/year). Faye (2010), for his part, superimposed curves showing changes in the position of high-water marks in the area from Ndar-Tout to the Hôtel Dior at the hydrobase, and calculated an average rate by linear regression of -0.20 m/year over the entire 1927-2005 period. Both early observations and recent episodes indicate rapid loss of large quantities of beach sand, recurrent marine submersions and house collapses at Goxummbac and Guet-Ndar (Sy *et al.,* 2010).

In the Saly area, Ba (1996) reports shoreline retreats of 10 to over 20 m between 1994 and 1996, corresponding to annual recession rates of 5 to over 10 m/year. According to Ibé *et al.* (1989), the average annual shoreline retreat between Joal and Sangomar is 2 m, while in the Nianing-Mbodiène sector, a study of the historical evolution of the coastline shows a retreat of 0.86 m, with a margin of error of 0.39 m.

In Casamance, Thior (2013) showed degradation of the coastal environment and socio-economic activities. The advancing sea has invaded the Cachouane cemetery and led to the destruction of certain tourist infrastructures (Lilibeu and the Gnikine camp). On the environmental front, Tall (2013), using multi-date satellite images, indicates degradation of the order of 2,417.8 ha between 1979 and 1986 and regeneration of 1,179 ha between 1986 and 2010.

Added to this are global (IPCC, 2007) and national (Gaye *et al.* 2008) forecasts of worsening climatic conditions, the consequences of which could be a rise in sea level, with non-negligible impacts as it affects several resources and economic activities.

Given the importance of the coastal zone, which is characterized by a high concentration of people and a wide range of socio-economic activities, the importance of protecting ecosystems that are vital for biodiversity must also be stressed. Over and above the benefits generated by key sectors such as fishing and tourism, which also depend on the environment being in good condition. What is not sufficiently emphasized is the fact that coastal zones are both subject to natural degradation and anthropogenic pressures. To effectively counter the impacts of coastal erosion, local coastal communities need to implement appropriate adaptation measures.

III. Research questions

For several years now, the coastline has been attracting growing interest from a multitude of stakeholders, who are expressing the need to act collectively on the natural and man-made processes likely to threaten the sustainable maintenance of environmental quality and the activities that take place there. Our research focuses on a number of issues.

Can we confirm from this study that the environment of the Commune of Kafountine is really threatened by coastal erosion? If so, doesn't coastal erosion lead to vegetation degradation and socio-economic damage?

Doesn't the advancing sea lead to the salinization of rice-growing land and groundwater? Doesn't this have an impact on rice growing and water quality?

Are the control strategies adopted by local communities and state and non-state structures appropriate? Should these strategies be strengthened?

These are the questions that guide our study and research, which is intended as a modest contribution to understanding the dynamics underway in the Commune of Kafountine.

IV. Research objectives

The main objective of this study is to analyze coastal erosion and its impacts in the Commune of Kafountine.

To achieve this goal, we have set ourselves three specific objectives:

- analyze the factors of coastal erosion in the Commune of Kafountine ;
- assess the environmental and socio-economic impacts of coastal erosion in the Commune of Kafountine;
- analyze adaptation strategies implemented by local populations in the face of coastal erosion in Kafountine Commune

V. Research hypotheses

Coastal erosion is the result of both natural and man-made factors, with both environmental and socio-economic consequences.

The work revolves around three specific hypotheses:

- the increase in coastal erosion is due to climate change and rising sea levels;
- this coastal erosion threatens the coastal environment as well as agricultural and tourist activities;
- the strategies developed to combat coastal erosion have limitations and need to be improved.
- **I. Conceptual analysis**

Coastal erosion :

Erosion itself is defined by Brunet *et al.* (2005) as "the removal of material from soils and rocks by natural agents: flowing water (torrential river erosion), glaciers and their melting (glacial erosion), wind (wind erosion), the sea (marine or coastal erosion), as well as rockfalls and landslides". Boudjéra (2010) defines coastal erosion as "a natural phenomenon, commonly observed and essentially characteristic of the superficial evolution of the earth's crust due to the displacement of sediments by waves. It is mainly due to a shortage of sediment. This shortage, which began to manifest itself at the end of the post-glacial transgression, has been accentuated in contemporary times by anthropogenic actions, in particular the construction of dams on rivers, the artificialization of coastlines, deforestation, etc. George and Verger (2006) see it as "a set of phenomena external to the earth's crust (or exogenous phenomena) that contribute to modifying the forms created by endogenous phenomena (endogenous tectonics and volcanism). This modification takes place through the removal of material: erosion proper".

A distinction can be made between long-term coastal erosion phenomena (multi-decadal scale), which result in coastal retreat, and short-term, seasonal coastal erosion phenomena, which are temporary because they are generally counterbalanced by periods of beach nourishment (Niang-Diop, 1995). This definition is in line with that of Paskoff (1983), who addressed the issue of coastal degradation by clarifying the dynamics of beach retreat and foreshore nourishment. According to Blivi (2001), coastal erosion marks beaches, a process that links coastal oceanographic parameters, sediment volume, shoreline configuration and anthropogenic factors. The European Commission (2004) sees coastal erosion as a progressive process of soil destruction by the sea.

We note that these definitions of coastal erosion are the result of a combination of natural and man-made factors that cause profound disturbances to the shoreline and coastal ecosystems. These disturbances act together to place beaches in a precarious situation.

Impact:

According to the Encyclopedia dictionary (2004 edition), an impact is a "*consequence, a slowing down of a strong action*"; indeed, the word impact is a loan from the Latin *impactum, supin of impingere* "to strike against", "to throw against", derived from *pangere* "to drive in", which is linked to an Indo-European root "to drive in, to fix" (André, 1999).

An impact can also be defined as the effect, over a given time and space, of a human activity on a component of the environment taken in the broadest sense of the term (i.e. encompassing biophysical and human aspects), compared with the likely situation if the project were not carried out (Wathern, 1998).

Coastline :

The coastline, in the physical sense of the term, is the strip between the lowest and highest seas, i.e. what the sea covers and uncovers (Akadje, 2010). For Klein (2003), it is a limited and often fragile space. La Rousse dictionary (2012) defines the littoral as a sinuous zone where contact is established between the sea or a lake and the land (the term has a broader meaning than rivage and côte, which respectively designate coastal areas subject directly or indirectly to the action of the sea).

The coastal zone is the contact zone between the sea and the land, two almost opposite environments: the land, solid, with landscapes, shapes and colors as varied as possible. It's an environment where most changes take place on a geological scale, except when man takes it upon himself to modify everything.

Strategy:

Lacoste (2003) defines strategy as part of the art of warfare, which consists of evaluating an army's ability to face the enemy on a relatively vast territory or theater of operations, by implementing a variety of tactics. For Chandler (1960), it involves determining the fundamental long-term objectives and goals of an organization, and then choosing the modes of action and resource allocation that will achieve these goals and objectives. In other words, it means putting in place the actions and allocating the resources needed to achieve these goals. It consists in defining coherent actions that follow a sequential logic to achieve one or more objectives. It is then translated, at operational level, into action plans by area and by period, possibly including alternative plans that can be used in the event of events that significantly change the situation. These definitions concur with those of Lusseault (2001), who see it as the art of organizing and coordinating a set of operations to achieve a goal.

We define the concept of strategy in relation to coastal erosion as the development of an action (or actions) with the aim of adapting to coastal erosion and its environmental and socio-economic impacts.

Having set out our theoretical framework, we can now turn to our research methodology.

CHAPTER 2: RESEARCH METHODOLOGY

Scientific research relies on the choice of a good methodology, which allows any scientist to verify the reliability or veracity of the study. In this work, the choice of methodology is structured in three parts: literature review, field data collection and processing, and spatial data processing.

I. Documentary review

It was carried out in the central libraries and documentation centers of the Geography Departments of Assane Seck University and Cheikh Anta Diop University in Dakar. We also visited the documentation databases of IRD, ISRA-CRODT and on the web. Finally, we visited the NGO Justice et Développement, the Aire Marine Protégée d'Abéné(AMPA) and the Services des Eaux et Forêts.

We consulted general works on the physical and socio-economic aspects of Casamance and the "Rivières du Sud" coastline, specialized works on the results of hydrodynamic agents (marine currents, swells, tides), articles, theses and dissertations related to our research theme. The information gathered enabled us to gain a better understanding of the causes and consequences of coastal erosion, as well as the strategies put in place to combat this phenomenon.

II. Field data collection and processing

2.1. . Field observations

Field observations enabled us to understand the evolution of the phenomenon under study. In the Commune of Kafountine, we observed the degradation of the coastal environment, the causes of accelerated coastal erosion, notably the removal of sea sand and logging, and the socio-economic impacts, such as the abandonment of rice-growing land.

2.2. . Interviews

We carried out individual and group interviews. The individual interviews involved the Mayor of the Commune of Kafountine, the Curator of the Abéné AMPA, the head of the Diouloulou Water and Forest Brigade and hotel owners. We discussed the environmental issues surrounding coastal erosion, in particular the factors, impacts and adaptation strategies.

Focus groups or group interviews were also organized in the eight(8) villages surveyed to gather information through focused discussions. The people targeted were senior citizens, to benefit from their wisdom, and grassroots community organizations. They enabled us to get a clearer picture of the start and reasons for the environmental dilapidation observed.

2.3. . Field surveys

A questionnaire was drawn up to carry out the surveys in the Commune of Kafountine, which comprises 19 villages.

Among these villages in the Commune, we selected eight (8) (Kafoutine, Abéné, Niomoune, Diogué, Hilol, Saloulou, Boko and Mantate) using the judgmental sampling technique[2] . The use of this technique is based on certain judgments about the population as a whole. The assumption underlying its use here is that we have selected (the pre-survey) the villages most likely to provide information on the impacts of coastal erosion and the adaptation strategies developed.

The Commune of Kafountine was made up of 1466 households in 2002 (ANSD, 2002). Based on this number, we determined a representative sample by dividing the parent population by the number of villages (n= N/8, i.e. 1466/8=183). This gave us 183 households across the eight (8) villages in the sample. It is important to calculate the probability of household selection. Since we know the sample size (n) and the parent population (N), calculating the probability of selection becomes a simple matter of division. This is obtained by dividing the sample size by the parent population (183 : 1466= 12.48). This means that each household in the targeted villages would have a 12% chance, or 1 in 12, of being selected.

How are these 183 households to be distributed among the eight villages?

We chose quota sampling[3] . We used the number of households per village as a basis for calculating the relative share (%) to be allocated per village. We adopted the following formula:

n n = Number of households per village/ Total number of households x 183 households

The sample reflects the percentage of households in each village.

Table 1: Number of households surveyed in the commune of Kafountine

Villages	*Number of households*	*Number of households surveyed*	*Percentage*
Kafountine	742	93	50,61
Abéné	323	40	22,03
Niomoune	181	22	12,35
Diogué	104	13	7,09
Hilol	54	7	3,68
Saloulou	37	5	2,52
Boko	16	2	1,09
Mantate	9	1	0,61
Total	1466	183	100%

2.4. . Survey data processing

Data collected during field surveys were processed using sphinx and Excel software. Graphical

[2] The crucial question in this case is objectivity: to what extent can we rely on our judgment to arrive at a typical sample? A. Diatta 2013

[3] amedé *et al,* (2013). research methodology, standards and writing techniques.

representations were made to assess the environmental and socio-economic impacts of coastal erosion. The graphs relate to the rate of vegetation mortality, rice paddy losses, decline in agricultural production and well water quality on the islands of Kafountine Commune.

III. Rainfall data collection and processing

To assess the impact of rainfall on the evolution of coastal erosion and its consequences, we sought to analyze the evolution of rainfall in the area. As the Commune of Kafountine has no rainfall station, we chose the one at Cap Skiring, which is less than a kilometer from the coast and closer to our study area. These data cover the period from 1978 to 2013.

These data were processed using Excel software, and enabled us to assess rainfall variability and its impact on the coastal zone of Kafountine Commune and Lower Casamance.

IV. Satellite data processing and diachronic mapping

IV.1 Spatial data used

The study of coastline evolution and land use is based on a remote sensing and Geographic Information System (GIS) approach. We used

four (4) satellite images acquired with the Landsat satellite in 1979, 1986, 2000 and 2015. The choice of these dates is linked to the rainfall variability noted in Basse Casamance, which remains an important element in the dynamics of the coastal environment. Indeed, the drought of the 1970s-1980s (Marius, 1986, Mahé, 2006) is said to have led to a sharp reduction in sediment supply from coastal river systems, with consequences for coastal areas (Niang-Diop, 1995 and Ndao, 2013). The study covers the years 1979 (the drought period) and 2015 (the current situation). The two intermediate years are 1986 (marked by the return of rainfall) and 2000 (anarchic occupation of space).

Table 2: Characteristics of the satellite data used

Satellite	*series*	*sensor*	*Acquisition date*	*Spatial resolution*
Landsat	L1	MSS	05-11-1979	57 m
Landsat	L5	TM	09-02-1986	30 m
Landsat	L7	ETM+	06-11-2000	30 m
Landsat	L8	OLI_TIRS	29/03/2015	30 m

Source : http://www.earthexplorer.gov

IV.2 Pre-processing and geometric correction of satellite images

Before undertaking the extraction of the vegetation boundary, we assessed the geometric quality of images acquired at different dates by superimposing them. Good conformity in terms of spatial resolution was noted between the ETM+ (2000), TM (1986) and OLI_TIRS (2015) images. On the other hand, images from the MSS sensor (1979) have a resolution of 57m; this does not allow them to be superimposed with the other images. So, since the images have different resolutions, we first resampled the 1979 images to a resolution of 30m. Then we georeferenced them using the 2015 image as a reference. These two steps made them superimposable.

IV.3. Mapping shoreline evolution and land use

Color composition and classification of satellite images

Using the satellite images selected, we produced color compositions, which are obtained by superimposing three spectral bands on the three primary color levels of red, green and blue. There are several types of color composition, including infrared false color, which was chosen for this study. The relevance of this color composition is that plants have high reflectance in the near infrared and absorption in the red. It is the best color composition for studying vegetation (Girard M.C and Girard C.M, 2010). It is obtained by combining spectral bands 7, 5 and 4 for the 1979 MSS image; bands 4, 3 and 2 for the 1986 TM and 2000 ETM+ images; and bands 5, 4 and 3 for the 2015 OLI_TIRS image.

Supervised or assisted classification has been used to group together spectrally similar areas on satellite images (Dieye, 2007). It requires prior knowledge of the terrain. The "Maximum Likelihood" classification algorithm is applied to all pixels in the image, based on the statistical parameters describing each spectral signature obtained. By comparing the values of each pixel with the spectral signatures of the various classes, the probability of the pixel belonging to each class is determined. The class with the highest probability is assigned to the pixel in question. We defined as many spectral classes (deep water, clear water, dense mangrove, clear mangrove, other vegetation, grassland, rice paddy, dry tanne, wet tanne, bare soil) as there are spectrally very different situations for each land use. The spectral signatures of each spectral class were examined in ENVI using training areas and separability analysis.

The results of this processing were vectorized and integrated as thematic layers in Arc GIS software to finalize maps of coastline evolution and land use.

Defining the reference line

The choice of reference line was made after careful examination of the images. Indeed, the spatial resolution of Landsat satellite images (30m for ETM+ and OLI images and 57m for MSS) is too coarse to measure the evolution of most coastlines. The only shapes whose evolution is compatible with these resolutions are spits and sandy islands located at the mouths of estuarine systems (which is the case of our study area). We have therefore chosen the vegetation limit as the coastline marker.

Indeed, according to some authors (Bird, 1985; Morton and Paine, 1985; Morton and Speed, 1998; Bird, 2000; Faye 2010), the seaward edge of dune vegetation (*Vegetation line, seaward edge of dune vegetation*) and the seaward edge of permanent*, stable, long-term vegetation* (*Line of permanent, stable, long-term vegetation*) can serve as valid reference points for the study of coastline evolution. In Kafountine Commune, the vegetation boundary is clearly visible on satellite images. In fact, dune vegetation is almost non-existent, and the somewhat coarse spatial resolution of the Landsat satellite images used makes it difficult to identify the high water mark. Digitization of the coastline has highlighted areas of erosion and accumulation for each date.

Defining thematic classes

For coastline mapping, we have merged all vegetation classes (mangrove, other vegetation, grassland, crop fields) into a *mangrove mudflats* thematic class, and the *continental dune*, *barrier beach*, *dry tanne*, *wet tanne and bare soil* classes into a *sandy spit* thematic class. The classes are merged and recoded into two themes: *mangrove mudflats* and *sandy spits*.

The results of these processes were vectorized and integrated as a thematic layer in a geographic information base in Arc Gis software, to format and finalize the diachronic cartography (figure 1).

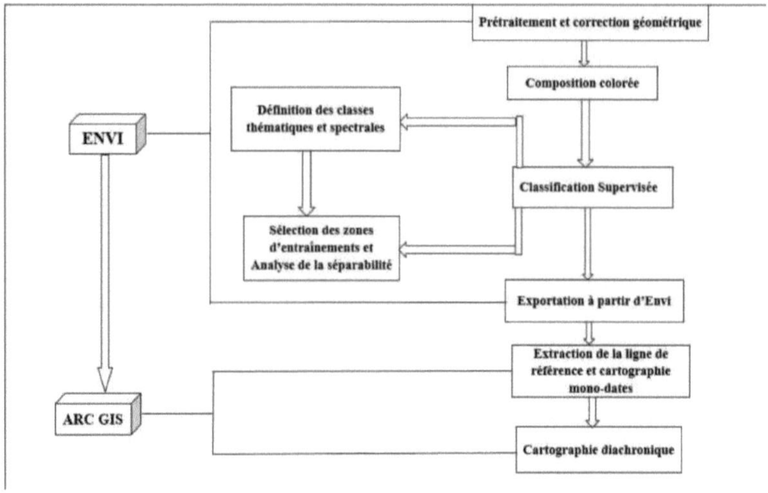

Figure 1: Summary diagram of coastline evolution and land use processing methodology.

Statistical method for calculating coastline kinematics

The statistical method for calculating the kinematics of the coastline in two dimensions (Erosion and Progradation) was carried out using Arc Gis software. It uses two known positions of the reference line (the oldest and the most recent). Polygons were created by digitizing the area between two coastlines. The polygons between these two coastline locations during the study period are calculated in metres to determine the average change, and divided by the number of years elapsed to obtain the annual change in kinematics. These linear distance measurements are complemented by surface area balances (in hectares), providing a two-dimensional view of the kinematics of the coastline.

Conclusion

In summary, this work is based on a theoretical corpus and a research methodology. The theoretical framework is presented along six axes: problematic, state of the art, research questions, objectives, hypotheses and conceptual analysis. It has enabled us to understand the functioning of coastlines in general and to define our research theme in particular.

Literature reviews, field data and satellite imagery were the main sources of information for this study. These data, collected and processed, enabled us to obtain results.

PART TWO :

CHARACTERISTICS OF THE SOIL ZONE AND EROSION FACTORS IN THE COMMUNE OF KAFOUNTINE

The coastal zone, in a broader sense than the coastline, determines the existence of a transitional environment between the ocean and the land. Two processes, both environmental and socio-economic, take place in the coastal fringe, but their manifestations are more visible on the shoreline.

This is a coastal zone where the influence of the sea has attracted people and their goods to settle, according to the following criteria: proximity and exposure to the action of the sea on the mainland, the climatic influence of the marine water mass, morphological and hydrological criteria built up by the blocking of water and sediments at the boundary of two marine and continental domains, economic and demographic criteria taking into account sea-related installations and activities (fisheries and tourism).

CHAPTER 1: CHARACTERISTICS OF THE KAFOUNTINE COASTAL ZONE

The coastal zone of the Commune of Kafountine has a number of physical characteristics, particularly compared with the rest of the Casamance natural region, since it abounds in mangrove estuaries with a very dense hydrographic network. As for the vegetation, it is marked by the presence of a variety of flora and fauna species that are part of the humid tropical zone of southern Senegal.

Its climate is coastal South Sudanian, with maritime influences and alternating trade winds and monsoons, depending on the season. The Kafountine coast is made up of two main areas: mangrove mudflats and dry land, which have been shaped over geological time.

1. Characteristics of the physical environment

Analysis of physical characteristics requires knowledge and control of the physical and anthropogenic processes at work in the environment. This approach makes it possible to target and understand the major physical dynamic components that control and modify coastal systems.

1.1. Location of the study area

The Commune of Kafountine is located on the southern coast of Senegal, more precisely in Basse Casamance, in the Ziguinchor region (Department of Bignona). It is bounded to the north by the Commune of Kataba 1, to the south by the Casamance River, to the east by the Diouloulou marigot and to the west by the Atlantic Ocean (map 1).

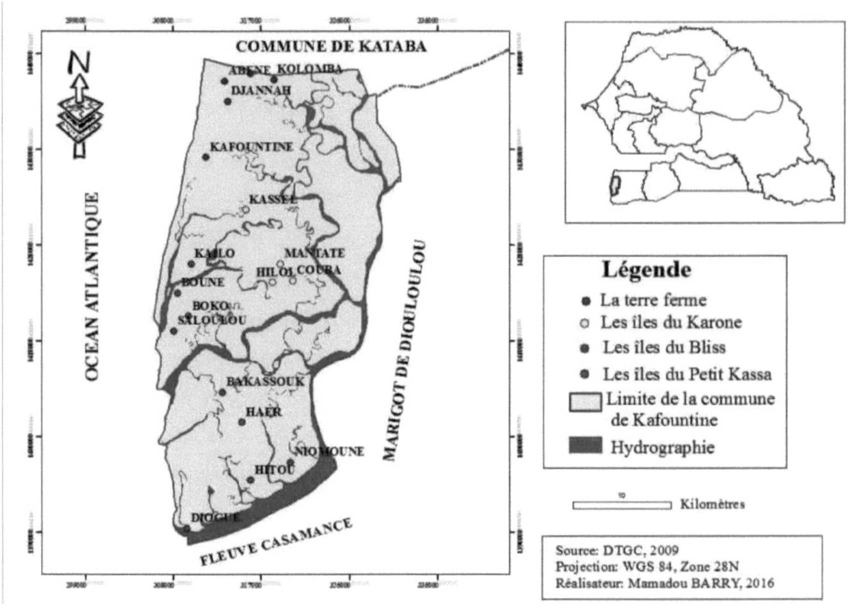

Cartel: Location of the Commune of Kafountine

The Commune of Kafountine comprises nineteen (19) villages covering a total area of 908 km^2 and is divided into two (2) distinct areas: the mainland (76 km^2) and the islands (832 km2)[4].

The mainland includes Kafountine, Abéné, Colomba, Diannah and Albadar. This is the mainland territory.

The Karone islands are Hilol, Couba, Coumbaloulou, Mantate and Kassel. This grouping represents the oldest settlements among the islands. This is the land of mangroves and gallery forests.

The Bliss islands are made up of the Kalissaye islands (Boune, Boko, Saloulou and Kailo). This is the area closest to the coast, and the villages have been built on the semi-fixed yellow dunes.

Petit Kassa, named after its proximity to Kassa. It encompasses the Diogué archipelago, comprising the villages of Diogué, Bakassouck, Haère, Hitou and Niomoune. [4]

The 2013 General Census of Population and Housing estimates the population of the Commune of Kafoutine at 30,071, representing a density of 33 inhabitants/km^2 . The population of Kafountine Commune is made up of 87% Diola-Karone, 10% Mandingo. The remaining 3% is made up of Peulhs, Manjacques, Mancagnes, Serrère etc. (PLD, 2009).

Its geographical position gives it the particular physical characteristics (sea tides, mangrove

[4] PLD, 2009

mudflats, vegetation, barrier beaches) found in coastal areas. This physical environment is associated with socio-economic activities (tourism, agriculture, etc.) that remain under threat from coastal erosion.

1.2. Geological and geomorphological overview

Like most coasts, the geological history of the "Rivières du Sud" coastline is essentially conditioned by eustatic variations and climatic changes (Diop, 1990). Recent Quaternary evolution plays an important role in the geological evolution of the Senegalese coastline in general and the Casamance basin in particular, especially in its estuarine part (Vieillefon, 1975; Diop, 1975; Kalck, 1978; Sall, 1982; Dacosta, 1989; Diop, 1990).

The Nouakchottian is a wet episode corresponding to the maximum of the Holocene transgression, around 5,500 years BP. The sea advanced into the interdunal corridors and river valleys up to an altitude of around +2m. This led to the formation of numerous coastal lagoons and marine gulfs.

Senegal's coasts were then regulated, with sea levels rising to over 3 m. All low-lying areas were transformed into gulfs. The fine sands and silts carried by the rivers were deposited in the form of azoic or fossiliferous terraces with *Anadara senelis, Tympanotomus fuscatus or Crassostrea gasar.* According to Kalck (1978), this period reached its peak during the Nouakchottian transgression, around 5,500 BP, when the sea invaded Casamance to form a wide gulf.

Around 4,000 years ago, the north-south longshore drift, which began around 4,500 years ago BP, led to the formation of barrier beaches that tended to close off the Gulf of Casamance. At Diouloulou, the base of the barrier beach is dated at 4310 years BP. Although this barrier beach does not outcrop, it does create a shoal that reduces marine energy and allows the lagoon character to develop behind it (Dacosta, 1989).

From the work of (Diop, 1975 and Sall, 1982), we understand the role of north-south or south-north longshore drifts. These drifts, carried by swells from either northwest or southwest, create highly mineralized sandy beaches, the oldest of which are located in the inner zones of estuaries. The addition of these thin, discontinuous sandbars and the filling in of the "inter-barrier" lagoons are at the origin of the formation of estuary island groups. This may explain the presence of islands in the Commune of Kafountine.

Around 3,000 years BP, the progress of the barrier beaches was such that the Gulf of Casamance took the form of a large lagoon where mangroves could develop intensively.

This is the case of the sandy spit on the bird peninsula in the Commune of Kafountine; a sandy spit built by the north-south longshore drift carried by north-westerly swells.

Towards the end of the 1500 BP period, Casamance took on an appearance comparable to that of today. The closure of the gulf by the barrier beaches led to a reduction in marine input and, consequently, a significant slowdown in morphological evolution. The low terraces defined by Vieillefon (1975) were probably individualized at this time.

Towards the end of the 1500 BP period, Casamance took on an appearance comparable to that of today. The closure of the gulf by the barrier beaches led to a reduction in marine input and, consequently, a significant slowdown in morphological evolution. The low clay terraces probably date from this period.

According to Kalck (1978), the filling of the Gulf of Casamance is the result of a combination of two progradation systems:

- the first corresponds to the development of the barrier beaches that close off the gulf; the larger group progresses in the north-south longshore drift, the smaller in the opposite direction, but both groups move progressively westwards,
- the second, determined by the first, corresponds to the advance of mangrove mudflats, which are gradually filling in the lagoons, to the great benefit of the Neolithic populations who lived in this area, leaving behind heaps of shells that show the importance of the consumption of numerous shellfish (mainly *Annadaras* and Oysters), attesting to the presence of mangroves (IDEE Casamance 2013).

Artificial piles of fossilized shells known as *Kjokkenmodding* can be found in the Commune of Kafountine.

These geological formations are the result of eustatic and climatic variations during recent Quaternary sedimentation. This coastal dynamic continued until 200 BP and led to the formation of morphological units. In the Commune of Kafountine, these formations are currently topped by aeolian dunes that line the coast. Two dune systems can be observed: living dunes on shell-rich beaches and semi-fixed dunes.

1.3. Hydrological overview

The Commune of Kafountine is drained by a multitude of small marigots (*bolongs*) that generally flow into the Diouloulou marigot or the Casamance River. These are Kafountine Boronda, Essoulou, Ebédou, Assouti Kassika, Balandiou, Kalissaye and the Rivière aux Huîtres (map 2).

They form a network of small streams that merge behind the spire of the bird peninsula in the form of a lagoon.

Kafountine Boronda is joined to the south by Essoulou, which, opposite the island of Boune, forms with Assouti Kasssika the confluence where the waters meet to flow into the ocean via Ebédou at a gap opened between the distal end of the bird peninsula and the Saloulou dunes.

The mouth of the Kalissaye is joined by the Rivière aux Huîtres to the south of the bird peninsula. These small streams, although fed by the Casamance River via the Diouloulou marigot, are subject to the influence of marine waters when they rise during low-water periods, especially in March-April during the low-water period of the Casamance River, when runoff remains very low (Diatta, 2012). These surface waters, which flow throughout the Casamance estuary, maintain the mangroves and animal species. They serve as a means of communication and enable the mobility of islanders (pirogue). The island populations practice artisanal fishing and fish farming in these waters, as well as in the *bolongs* (map2).

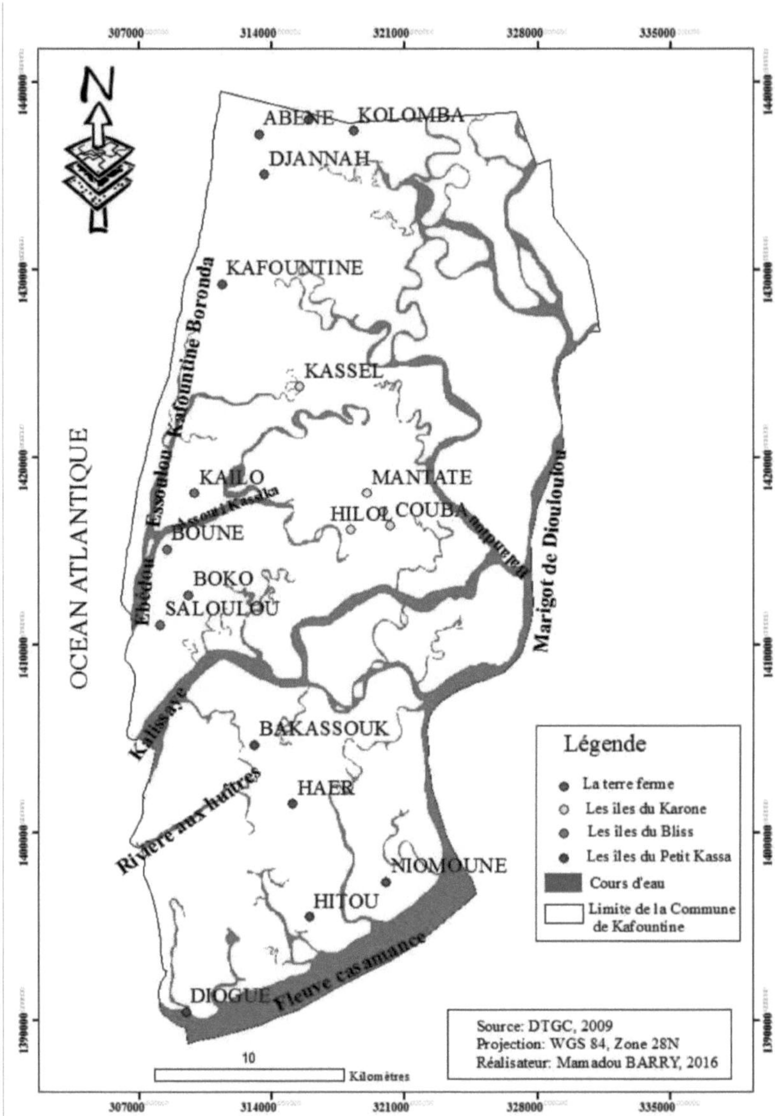

Map 2: Hydrographic network of the Commune of Kafountine.

I.4. Climate overview

The study of the geomorphology of a given region's coastline cannot be understood by ignoring the fundamental relationship between coastal forms and climate. The climate

is therefore a decisive factor in the development and evolution of coastal forms.

Climatically, the Commune of Kafountine belongs to the Sudano-Sudanese zone, and its coastal position places it in the coastal Sudano-Sudanese domain.

This affiliation is characterized by a study of the climatic parameters that distinguish it from the rest of Senegal.

Throughout the year, the Commune of Kafountine, as part of the South Sudanese coastal region, is subject to the successive influences of the trade winds and the monsoon. These two factors give rise to two clearly differentiated seasons: the dry season, driven by the trade winds, and the rainy season, characterized by the monsoon.

The dry season: winter circulation in Senegal, i.e. the period corresponding to boreal winter or the dry season, is essentially characterized by the circulation of the Alize. Depending on its trajectory and the ensuing consequences, the Alize is the flow that characterizes this season, with a dual behavior: it can be described as a maritime Alize or a continental Alize.

It runs from November to May. During this period, the Commune of Kafountine is traversed by the maritime trade wind, originating from the Azores anticyclone cell.

The rainy season: generally from May to October. This season, known as the wintering or summering period, stems from the Saint Helena anticyclone cell, which has the characteristics of a trade wind. As it crosses the equator, it is deflected and becomes a monsoon. In this case, the flow changes direction. For monsoon circulation to occur, the Meteorological Equator must be offset from the Geographical Equator. This south-westerly monsoon flow discharges its precipitation in line with the progression of the Meteorological Equator.

Its long oceanic journey loads it with moisture as it rises towards the north of Senegal. The water vapour contained in this air mass explains the formation of clouds, the increasing development of clouds and, logically, the drop in precipitation.

Analysis of rainfall (1978-2014) at the Cap Skirring station shows variability in rainfall contributions (figure 2).

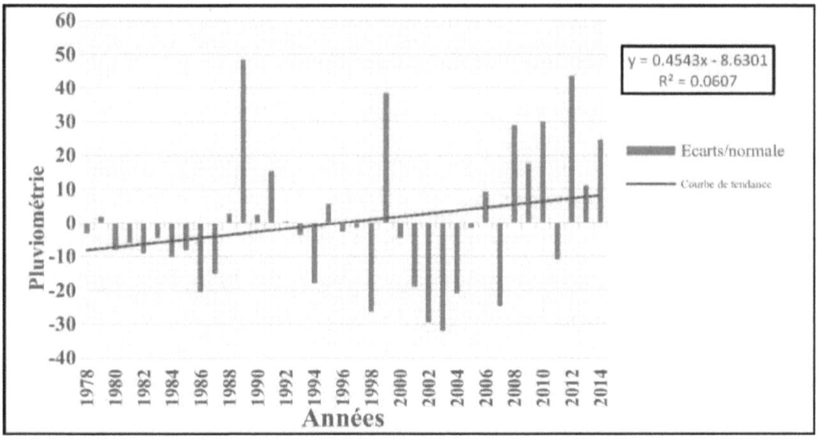

Figure 2: Rainfall trends from 1978 to 2014 at the Cap Skiring station (Source: ANACIM, 2015).

Generally speaking, this fluctuation is characterized by a predominance of deficit years at the expense of surplus years. Twenty-two (22) years are in deficit, compared with 15 years in excess of the average (1215.83 mm).

Rainfall trends can be divided into 3 periods:

the first period (1978 to 1992), the second period (1993 to 2005) and the third period (2006 to 2014).

The first period comprises fifteen (15) years, from 1978 to 1992. Within these 15 years, there are 9 deficit years and 6 surplus years. In percentage terms, the deficit years represent 60% and the surplus years 40%. The deficit years are 1978-1980-1981-1982-1983 to 1992. Among these deficit years, eight (8) are very slightly in deficit. These were (1978-19801981- 1982-1983-1984-1985 and 1987). These deficit years are a continuation of the drought of 1970. The surplus years were 1979-1988-19891990-1991 and 1992. Within these surplus years, we have five (5) very low surplus years, corresponding to 1979-1988-1990-1991 and 1992, and one medium surplus year, 1989(48.4).

The second period covers thirteen (13) years, eleven (11) of which were in deficit and two (2) in surplus. It covers the period from 1993 to 2005. It can be seen that 84.62% of this second period was in deficit, compared with 15.38% in surplus. The deficit years are: 1993-1994-1996-1997-1998-2000 to 2005. Within these deficit years, we observe seven (7) very low deficit years (1993-1994-1996-1997-2000-2001 and 2005) and four (4) low deficit years. Next, the surplus years can be summarized as two. These were 1995 and 1999. Within these years, 1995 can be characterized as very low surplus (5.53) and 1999 as low surplus (38.4). Finally, the second period shows a relatively negative trend. Nevertheless, we have two surplus years during this overall deficit period.

The third period comprises nine (9) years from 2006 to 2014. Within these nine (9) years, there are seven (7) surplus years and two (2) deficit years. In percentage terms, surplus years represent 77.78%, while deficit years account for 22.22%. The surplus years are: 2006-2008-2009-2010-20122013 and 2014; among these surplus years, there are three (3) very low surplus years (2006-2009-2013) and three (3) low surplus years (2008-2010-2014). Finally, 2012 stands out as a year with a moderate surplus (43.46).

The deficit years are 2007 and 2011. Within these deficit years, we have a very small deficit in 2011 (-10.86) and another small deficit in 2007 (-24.54).

It is interesting to note, however, that the third period follows a relatively positive trend, in contrast to the other periods, which are characterized by a largely negative trend.

An overall analysis of this series shows that rainfall is relatively deficient. This rainfall deficit has led to a reduction in the flow of the Casamance River, and freshwater inflows have become insufficient to compensate for the upwelling of sea water. As a result, the salt water tongue has risen all along the river, limiting the cultivation of land and making well water unsuitable for consumption.

Land use and occupancy

In the Commune of Kafountine, the vegetation is dominated by mangroves and palm groves. The main land uses are agriculture, tourism and fishing. The Commune's geostrategic position favors a diversity of activities, which are increasingly threatened by coastal erosion.

11.1. Plant cover

The vegetation cover of the Commune of Kafountine is dominated by mangrove vegetation alternated in places by other less representative species.

Mangrove is a wetland and brackish water formation located in halomorphic formations. It covers more than half of the Commune's coastal vegetation, dominated by the insular part commonly known as the mangrove zone. According to Cormier-Salem (1999), this is the zone that is directly subject to tidal influences, and thus corresponds to the supralittoral and meridional stages and to the upper part of the infralittoral stage in relation to the classic zonation system adopted in the marine environment.

The plant species of which it is composed are referred to as mangroves, three of which generally colonize this zone. Mangroves of the genus *Rhizophora*, *Avicennia* and *Laguncularia* are generally found here.

On the plateau, we find a dense deciduous forest characterized by slender shafts. Plant formations include *Dialium guineensis* (Tamarind), *Khaya senegalensis* (Caicédrat), *Saba senegalensis (*Mad), *Borassus aethiopum* (Rônier), *Ceiba pentendra* (Fromager), *Andansonia digita* (Baobab), *Pterocarpus erinaceus* (Ven) and others.

The slopes and lowlands are mainly planted with palm groves dominated by oil palm (*Elaeis guineensis*). These palm groves form a homogeneous stand at the edge of upland forests, on the edges of villages and on slopes, depressions and valleys. This vegetation reveals a certain floristic richness, which can be linked to the presence of the water table at a relatively shallow depth.

11.2. Socio-economic activities

The landscape of the Commune of Kafountine is a mosaic of rich and dynamic resources. It includes large areas of forest of diverse composition, coastal areas and stretches of freshwater, providing important opportunities for socio-economic activities on which the Commune's economic development depends (figure 3).

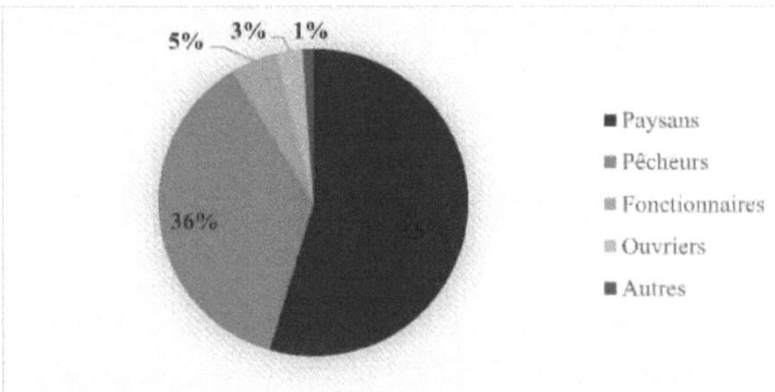

Figure 3: Population occupations (Source: Survey data, 2016)

Agriculture accounts for almost all of the Commune's population and provides the main food and cash resources. It is made up of rice growing and cash crops, essentially arboriculture, cashew growing and market gardening. Fishing is an income-generating activity in the Commune of Kafountine. In general, the beneficiaries of this sector are foreigners from the north of the country (*Lebous* and *Guet Ndariens* fishermen) and a minority of Ghanaians and Burkinabés who fish on the high seas.

As for the natives, they fish in the small mangrove *bolongs*. Sedentary fishing is most active in the dry season. In the winter months, it continues to be practiced to satisfy both nutritional needs and the requirements of the cash economy. The fishing calendar is therefore designed to allow simultaneous fishing and farming.

Tourism is also a well-established activity in the Commune of Kafountine. It generates and provides more income and jobs. All along the coastal strip from Abéné to Kafountine, the territory is occupied by tourist infrastructures and permanent residences. Nowadays, other forms of tourism are being sought: integrated tourism, ecotourism and beach tourism. Despite its role in the local economy, it is under threat from coastal erosion and faces customer problems linked to the armed conflict in Casamance and a lack of organization.

Conclusion

The coastal zone of the Commune of Kafountine is made up of three units with a South Sudanian coastal climate favorable to forests. It is drained by multitudes of watercourses flowing into the Diouloulou marigot and the Casamance River, and are surrounded by mangrove vegetation. Fishing, agriculture, trade and tourism are the main economic activities pursued by both local and foreign populations, most of whom come from neighboring countries. These physical and socio-economic characteristics make the coastal zone generally vulnerable.

CHAPTER 2: COASTAL EROSION FACTORS IN THE COMMUNE OF KAFOUNTINE

The coastline is considered to be a clean break, the line along which sea and land meet. This linear interference, due to its instability in the short term (tidal action) and/or in the medium or long term (sea-level variations), is a sign of insufficiency and imprecision, and therefore ambiguity. In the broadest sense, the coastline contains marine and coastal ecosystems that influence economic and social activities. As a result, the entire Commune of Kafountine can be described as a coastal zone.

Subject to the influence of complex natural and anthropogenic phenomena, the Commune of Kafountine is significantly affected and threatened by coastal erosion.

Coastal erosion is mainly due to natural factors, mainly marine (swells, tides, currents) and subaerial (wind, rainfall, etc.). These interact and contribute to the erosion of the physical coastal environment.

Anthropogenic action also plays a key role. Indeed, human activities interfere with natural agents, thus contributing to the worsening of coastal erosion. Their action can be direct, as in the removal of sand from beaches. It can also be indirect, through the artificialization of the coastline, which modifies coastal sediment transfers.

1. Natural factors

The natural agents involved in modifying the coastline of the Commune of Kafountine are typically marine and underwater.

1.1. Marine agents

1.1.1. Swells and waves

Swell is defined as the orbital undulatory motion of water molecules, whose orbits are damped with depth by internal friction, under the influence of the wind. It plays an important role in the dynamic behavior of beaches (Soumaré, 1996).

Studies carried out on the continental shelf by various authors (Sall, 1982; Diop, 1990; Niang-Diop, 1995; Faye, 1993) confirm that there are two main types of swell in this area:

The north-westerly swell: originating from the North Atlantic, it is present throughout the year and arrives directly on the northern coast. According to Guilcher (1954) in Diop (1990), it is classified as a long swell, and is particularly sensitive in the northern part of Senegal (from the Saloum to Casamance). It undergoes refraction and a series of diffraction phenomena at at least three points: the Pointe des Almadies, Cap Manuel and Pointe de Bel-Air (Riffault, 1980 in Niang-Diop, 1995), and arrives in a southwesterly direction, losing all or part of its energy. These changes in the swell result in its damping before it reaches the southern coast.

In the Commune of Kafountine, the effect of its direction from north to south presents a simple morphology, with examples of a sandy spit on the Presqu'île aux oiseaux, small scattered "spits" and the recent sandy spits of Diogué.

Southwest swell: originating in the southern Atlantic, this swell only affects the Petite Côte. According to Sall (1982), the influence of the south-westerly swell in this sector of the coast is remarkable up to northern latitudes, and its effects on morphogenesis are notable in the rainy season. In the Commune of Kafountine, its influence is most noticeable in the orientation of sandy dunes from south to north, conveyed by longshore drift.

As it approaches the coast of Kafountine, the characteristics of the swell change, its camber amplifies and it breaks: it is the wave that influences the bottom and acts on sediment transport. Waves, swells and currents are the most important forces influencing and determining the dynamic behavior of beaches (Ibe and Quelennec, 1989). These factors depend on the wind, its strength, direction and the distance over which it blows.

The combined action of swells and waves (very strong during the trade winds) causes a major transport of sand during the spring equinox (March-April). This results in a major modification of the beach profile (Soumaré, 1996), with part of the sediment being carried along by coastal drift.

1.1.2. The tide

The tide is considered to be a movement of rising or falling water periodically observed at sea. In particular, it is the difference between high tide and low tide, known as tidal range.

In fact, it is the result of the action of celestial attractive forces such as the moon and the sun on the water masses (Diop 1990; Niang-Diop, 1995; Adjoussi, 2001).

It manifests itself in two ways:

- on the one hand, when the sun and moon are in attraction (i.e. aligned), the amplitude of the tides (the difference in height between high and low water) becomes greater: this is known as a spring tide;
- on the other hand, when the sun and moon are in opposition (at right angles to each other), tidal amplitude remains low: this is called neap tide.

At the mouth of the Casamance River, the amplitude is 1.40 m during spring tides and 0.90 m during neap tides (Soumaré, 1992). The tides play a significant role in the distribution of sediments on the beach, through the highly erosive currents to which they can give rise (Domain, 1976; Rebert, 1983; Diara 1999). In winter, rainwater combined with high tides causes damage to the islands of Kafountine Commune.

I.1.3. Coastal drift

Longshore drift is a current that runs parallel to the coast. It is induced by waves arriving obliquely to the shore. Littoral drift is the main cause of sediment transit along the coast (Diop 1990; Niang-Diop, 1995; Diara, 1999; Sy, 2013).

The Casamance estuary lies at the point where the north-south drift from the North Atlantic meets the south-north drift from the southern swells, reinforced in winter by the south-west monsoon (Pélissier, 1966). These two opposing drifts attenuate each other's effect and, in places, cause the coast to silt up. The role of these drift currents, which are generally directed from north to south, is evidenced by the presence of south-facing sandy spits along the Senegalese coast (langue de barbarie, Sangomar, etc.) (Diop, 1990; Diaw 1997; Sy *et al.*, 2010; Faye, 2010; Sy, 2013).

In the Commune of Kafountine, for example, longshore drift is responsible for the construction of the coastal spit of the bird peninsula, as well as smaller spits such as those at the tip of Kalissaye and to the immediate south of the mouth of Kalissaye. In general, these spits migrate southwards and remain under the influence of longshore drift.

I.1.4. Currents

Currents are dependent on atmospheric pressure on water masses. They are veins of currents perpendicular to the shoreline, through which the flow of water takes place.

the seaward movement of surface water pushed towards the coast by waves and wind (Domain, 1976; Rebert, 1983; Faye 1993).

They have a remarkable influence on the beaches of open coasts such as that of the Commune of Kafountine. They play an important role in the sedimentation process and influence temperature distribution in marine areas, as well as the distribution of benthos and plankton (Guilcher, 1954 in Soumaré, 1996). According to Diallo 1982, there are two main currents along the coast:

<u>The January or winter current:</u> under the action of the trade winds, the surface layer of warm Liberian waters flows south-westwards, causing a compensatory upwelling of 50 to 100 meters of semi-deep cold water (Demoulin, 1967). The entire Senegalese coast is then bathed in upwelling, with temperatures ranging from 16°C to 17°C at the surface and 15°C at depth. These upwellings are of two types: a weak upwelling in December-January and a strong one in February (Faye, 1993).

<u>The July or summer current:</u> waters driven back by the equatorial countercurrent in the Gulf of Guinea sector begin to drift northwards at the surface against the African coast. "The warm, desalinated waters are directly influenced by the equatorial calms and the more or less strong west and south-west monsoon winds.

From November to May, this surface current gives way to an internal undercurrent. Here too, there are two types of current: a stable current between July and August, and an unstable current between September and October.

1.1.5. Sea level

Sea-level rise depends on a complex set of factors that interact on different time and space scales (Emery and Aubrey, 1991 in Niang- Diop, 1995). Variations on a global (eustatic) scale depend essentially on the volumes of the ocean container and the quantity of ocean water (Pirazzoli, 1993).

A few decades ago, the international scientific community was sounding the alarm about the impact of human activities on the climate balance. Today, recent observations and measurements presented in the latest reports of the Intergovernmental Panel on Climate Change (IPCC) leave no room for doubt that the planet's climate is changing. Most of the changes observed over the last fifty years are attributable to human activity. This was clearly stated in the IPCC's fourth report (IPCC, 2007), which stipulates that

"Most of the observed increase in global average temperature since the mid-20th century is very likely due to the observed increase in anthropogenic greenhouse gases".

The causes of sea level variations result essentially from two factors (Niang-Diop, 1995, Abe, 2005; Niazi, 2007):

- variations in ocean volume, caused by changes in sea temperature and salinity;
- variations in the mass of water contained in the ocean, mainly as a result of exchanges with continental reservoirs, mountain glaciers and polar ice caps.

These two factors are profoundly linked to rising temperatures. Thermal expansion of the oceans linked to sea warming is responsible for two-thirds of the current rise in sea level, with the remaining third resulting from melting ice.

Like all bodies, water becomes denser and expands when subjected to temperature. This rise in temperature is partly linked to the ever-increasing concentration of CO2 in the atmosphere. As a carbon stock, the ocean releases CO2 when its waters warm up. The quantity of CO2 in space is now such that the atmosphere and plants can no longer recycle it (Niazi, 2007). This high level of carbon dioxide in space is increasing the Earth's temperature, resulting in the melting of the Antarctic polar ice caps.

Global warming has more than just physical causes. Man is also partly responsible (Niang-Diop, 1995; Niazi, 2007; IPCC, 2007; Boudjera, 2010; Thiam and Crowley, 2014). Economic activities based on the consumption of fossil fuels (oil, coal, etc.) release substances into the atmosphere that are too harmful to the ozone layer that protects us from excess harmful solar rays. Today, this layer is threatened by the emissions produced by human economic activities. These emissions are mainly composed of carbon dioxide (CO_2), methane (CH_4), nitrous oxide (N_2O) and chlorofluorocarbons (CFCs).

These gases absorb infrared radiation. They are likely to increase the greenhouse effect and warm the lower atmosphere (the troposphere), triggering increased evaporation and amplifying the role of water vapour. For example, the IPCC (1995) forecasts a rise of 0.45 m by 2100 (Houghton *et al.*, 1990), while the OECD (1993), at the Second International Climate Conference, predicted a rise of 0.3 to 0.5 m by 2050.

The serious consequences of rising sea levels can be deduced from the work of Bruun (1962), who showed that a 0.3 m rise in sea level can lead to coastal retreats of over 33 m, with even greater potential for regression in low-slope marsh areas (Ibe and Quennelec, 1989). In addition, studies carried out on the vulnerability of Senegal's coastal zones to climate change (Dennis *et al*, 1995) have shown that these rates of sea-level rise could lead to accelerated coastal erosion, flooding of low-lying coastal areas (mangrove estuaries in particular) and increased salinization of soils and surface and groundwater.

In the Commune of Kafountine, extrapolations suggest that global sea-level rise may indirectly affect the coastal landscape.

1.2.1. Subaerial agents

Subaerial agents are also responsible for coastal degradation. However, they are less erosive than marine agents. They include wind and rain.

1.2.2. The wind

Wind is a highly effective erosive agent, acting on soils essentially by deflation (Seck, 2001). On the coastline of Kafountine Commune, the wind erosion mechanism is generally negligible, and the effect of the wind is normally slowed by a curtain of filao. On the other hand, wind action is very evident on some coastal dunes, which are more or less vegetated.

Wind can increase or decrease shoreline drift. They can also increase wave camber. Indeed, if the wind blows in the opposite direction to the waves, it slows them down, and if it blows in the same direction, the waves camber more. In this case, the wind can play a role in reducing the size of the beach.

1.2.3. Rain erosion

The splash effect and runoff are two examples of rain erosion.

Erosion caused by the impact of raindrops is known as splash or battance (Seck, 2001). As raindrops fall on bare soil, they break up into fine droplets that rebound after becoming loaded with silt, clay and mineral salts. The raindrops create a tangential tearing force, proportional to their kinetic energy, breaking up clods and aggregates and ejecting the torn particles. Following the sorting of elements by the splash, the finest grains insinuate themselves into the pores of the soil, between the thick materials, where they form a kind of clay cement that hardens strongly as it dries out (Rougerie, 1954 in Sall, 1971).

In some cases, the surface layer of the soil forms a compacted layer, clogged by the beating of the rain, known as the "pellicule de battance" or "couche de battance" (Dupriez and De leener, 1990).

What's more, the effectiveness of battance is strongly linked to the nature of the soil and the topographical slope. The more clay and silt the soil contains, the more compact and impermeable the capping film will be.

As a result, the soil surface is more conducive to runoff than infiltration.

Runoff is the phenomenon of water flowing over the surface of soils during heavy rainfall. This leads to landslides or slumps in the barrier beaches, which feed sediment to the low-lying beaches and accelerate the retreat of the high-lying beaches. Along the Kafountine coastline, we have observed that sediments detached from the micro-cliffs are rapidly dragged along by the waves.

These natural factors of coastal erosion do not act in isolation. They combine and complement anthropogenic factors to accelerate and intensify coastal erosion.

11. Anthropogenic factors

The natural factors described above cannot be held solely responsible for the mobility of the coastline. Man, through his many uses and occupations of the coastal zone, contributes directly or indirectly to its erosion.

In the Commune of Kafountine, many socio-economic activities have been developed, leading to an imbalance in the coastal system.

11.1. Coastal development

During our field visits, we observed the sometimes anarchic encroachment of various tourist and fishing developments in the coastal zone.

11.1.1. Poor tourist facilities

Tourism is a well-established activity in the Commune of Kafountine, generating and providing more income and jobs. Indeed, Commune Kafountine is one of the most attractive areas in Basse Casamance, thanks to the many forms of tourism that have been developed.

These include integrated rural tourism or value tourism. According to Diatta (2007), this is a form of tourism developed by local people, based on the promotion of traditional cultures. This form of tourism brings local people closer to tourists, and has enabled some of them to invest in tourism by building a receptive business. Other forms of tourism are also in demand, such as ecotourism, which aims to make the most of nature, landscapes and flora and fauna. This form of tourism is based on the Commune's relatively dense vegetation, with a wealth of flora and fauna, marked by the predominance of the mangrove ecosystem. Finally, seaside tourism plays an important role. The calm, little-frequented Kafountine sea attracts tourists in search of relaxation.

These are the main factors behind today's flourishing tourist infrastructure. Hotels and camps have sprung up all along the Commune's coastal zone, particularly in the coastal sectors of Abéné, Dianna and Kafountine. These include the Kalissaye, Karone, La Nature and Atlantic Club hotels. These tourist developments were built on the coastal dunes adjacent to the beaches, and sometimes even on the upper beach (map 3). The construction of housing directly on the barrier beach has the effect of fixing sediments and preventing sediment exchange between the top (sediment stock) and bottom of the beach. As a result, erosion is inevitable in exposed areas.

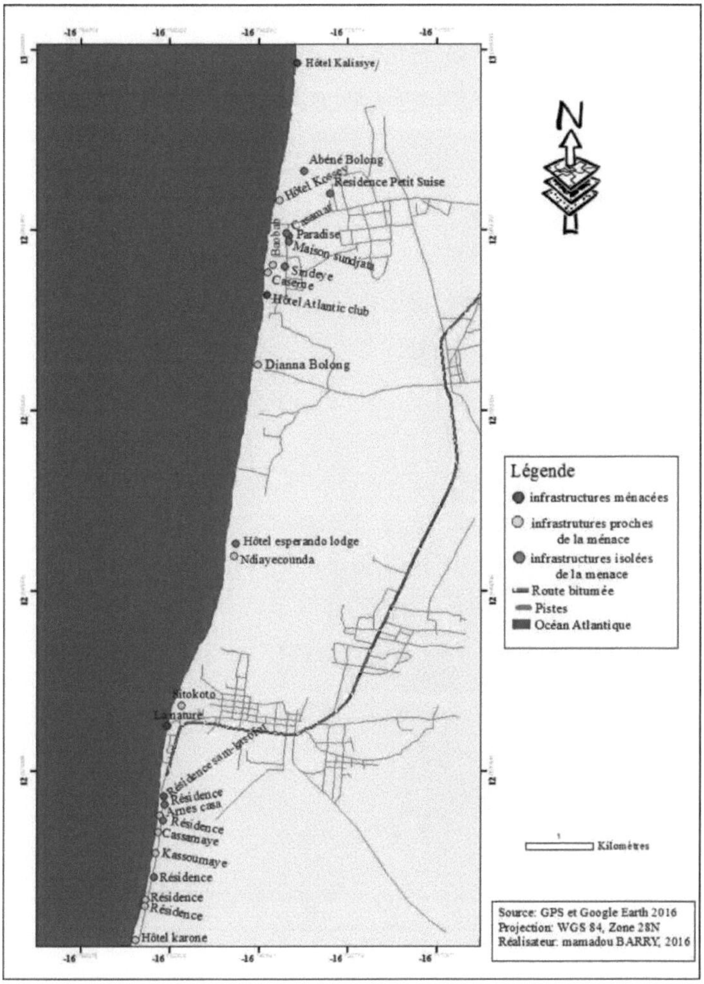

Map 3: Tourism infrastructure along the coast from Abéné to Kafountine

We can also point out the four-wheeled trucks or motorcycles on the beach that transport tourists in search of the beautiful landscape of dunes and greenery. These vehicles, which travel at high tide, destroy the barrier beaches and the vegetation that grows along them.

11.1.2. Problems related to fishing developments

Fishing is one of the most prosperous economic activities, and one of the main sources of direct and indirect income for the Commune's population. It is practiced from

permanently by local people and seasonally by foreigners from northern Senegal (*Lébous* fishermen and *Guet Ndariens*) and neighboring countries (Burkinabe, Ghanaians, Malians, etc.).

The dynamism of the industry in the Commune of Kafountine has encouraged the development of ancillary activities such as fish and other seafood processing and fish trade. These activities have led to intense use of the sandy spit in the coastal areas of Kafountine, Abéné and Diogué (map 4).

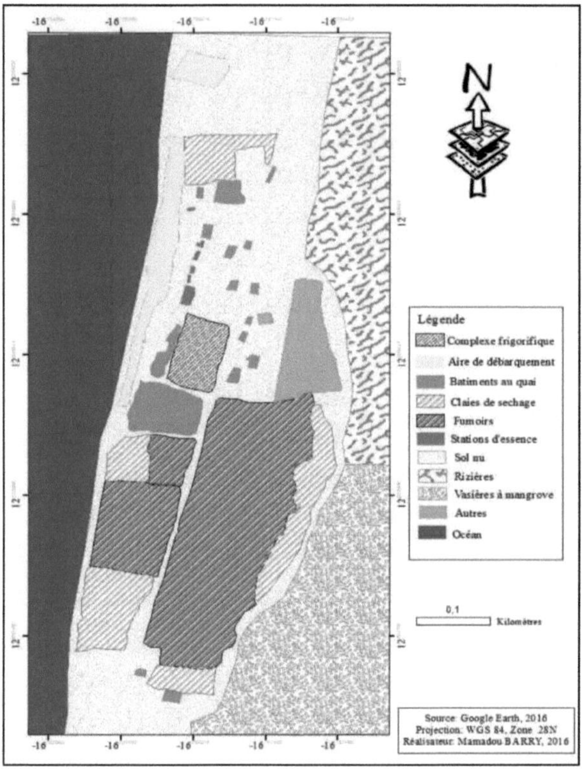

Map 4: Fishing infrastructure installed in the Kafountine fishing quay

The vegetation that fixes the sand in the sandbars and dunes is thus trampled underfoot. Series of shelves, arranged in parallel strips, run transversely along the shore in the upstream part of the foreshore. These are a series of drying racks, forming an esplanade from the high beach upwards, and in some places spread out to create walkways. The trellises established on the foreshore are used to dry fish catches (Photo 1).

Photo 1: Claies de séchages sur la plage à Kafountine (Barry, March 2016)

At this level of the beach, plank huts are set up on the low-strength sand to provide shelter for fishmongers, fishermen and processors. As a result, net residues discarded by fishermen all over the beaches are a source of coastal pollution. This net debris contributes to the degradation of coastal ecosystems (seagrass beds, reefs), which play an important role in maintaining the balance of the coastal environment.

There are also a number of fishing-related economic buildings. These include petrol stations (Star énergies, Omega petroleum, Société Africaine de Pétrole, ORYX énergies, Total), which supply fuel to motorized pirogues. These stations are part of the list of infrastructures built on the sandy spit (Photo 2).

Photo 2: Total petrol station installed in the fishing quay at Kafountine (Barry, Mars 2016)

Because of its role as an economic hub, this coastal area is subject to a sustained rate of anthropization. As a result, the coastline of the Commune of Kafountine is changing, with no less significant effects.

11.2. Sea sand harvesting

Man's actions on coastlines and beaches contribute directly to erosion. This contribution can also be seen in the extraction of sea sand, which is often preferred for building houses and for filling in.

Its appeal lies in its proximity, lower cost and particle size composition, which varies between 0.063 mm and 0.6 mm, with an average of 0.2 mm (Adjoussi, 2001). In the Commune of Kafountine, these samples were used for house construction and backfilling.

On the beaches of Kafountine and Abéné, for example, residential buildings and fishermen's houses can be seen just a few metres from the shore. These types of construction reflect the extraction and use of beach sand.

What's more, on the islands, sea sand is used for most of the house construction inside the village or on the beach (Kailo, Saloulou, Boko, Boune). In these villages, which were built on semi-fixed yellow dunes, traditional straw and wooden houses are beginning to disappear (figure 4), in favor of modern, hard-built houses, which often require considerable quantities of sand to be extracted. In fact, of the 183 households surveyed, only 6.56% were traditionally built (figure 4).

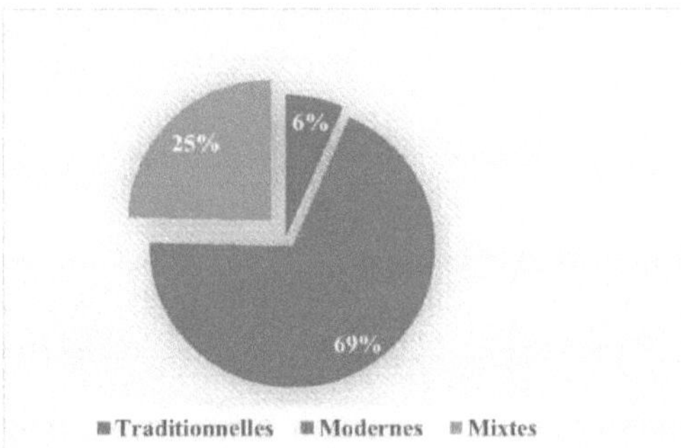

Figure 4: House types encountered in the Commune of Kafountine (Source: Survey data, 2016).

In the Commune of Kafountine, illegal removal of sea sand is a recurring problem. These sediments are used to make bricks for building houses, as well as traditional dikes to protect tourist infrastructures and mitigate the effects of run-off water.

II.3. Deforestation

Surveys carried out among local populations show that deforestation has reached a climax in recent decades. Indeed, forest resources have been subjected to excessive degradation by local populations and foreigners, particularly Gambians and Ghanaians.

According to the commander of the Aire Marine Protégée d'Abéné (AMPA), Mr Mamadou Sidibé, numerous illegal logging units supply the Gambian market. Despite this evidence, no major initiatives have been taken to protect the Commune's forest resources. Localities such as Dianna, Diana Kabar, Abéné, Albadar and Colomba supply Kafountine and surrounding areas (Kataba and Diouloulou) with firewood, timber and wood for handicrafts.

The Kafountine fishing wharf is an important site for unloading and smoking fish. Trees and shrubs are cut from the coast and in the forests to serve as fuel for smoking and shipyards (Photos 3). Thus, according to Diabang (2016), during the low season, an average of 3 trucks, 36 vans and 35 carts supply women fish processors every day. The price of a cart load varies between 10,000 and 20,000 Fcfa during periods of high catches, compared with 7,500 to 15,000 Fcfa during average catches.

Photo 3: Artisanal shipyard (A) and firewood for smoking fish (B) in Kafountine (Barry, December 2015 and March 2016).

The vegetation, which acts as a shield to protect the soil, is subject to heavy deforestation, which accentuates rainwater runoff. On the ground, this results in the digging of gullies along the slopes and breaches that contribute to beach subsidence, with the release of large quantities of sediment into the sea.

Although prohibited by the national forestry code, with the exception of dead wood, the right to use which is granted to local residents, the exploitation of mangrove wood (Photo 4) remains a reality in the Commune of Kafountine. Many villagers cut mangrove wood for various purposes. This sometimes irrational assault on the mangrove ecosystem is not without consequences for the future of the local population and the socio-economic activities that depend on it.

Photo 4: Stocks of mangrove wood in Boko (Barry, April 2016)

Mangrove forests fulfil important functions in terms of coastal protection, biodiversity conservation, the provision of woody and non-woody forest products, habitat, spawning grounds and nutrients for a variety of fish and shellfish, salt production and so on. Thanks to their intertwined roots, they also play a role in fixing and protecting loose soil, thereby protecting the coast against erosion. The complex three-dimensional structure of a 200 m strip of mangrove branches, trunks and roots can absorb up to 75% of the energy produced by wind-driven waves (Sarr, 2012).

Conclusion

Coastal erosion is a natural process that has always existed. Natural factors such as variations in sea level, together with marine agents (swell, currents, drift, etc.), all contribute to coastal erosion.

tides, etc.) and subaerial forces (precipitation and wind) have led to an imbalance in the coastal system. The effects of these natural agents on coastal evolution are exacerbated by human activities along the coast. The high concentration of human activity has led to a growing need for infrastructure (tourism and fishing) and building materials (sea sand), but has also made the coast more susceptible to coastal erosion. These various human activities have interfered with the natural process by reducing sediment supply, and have contributed to accentuating coastal erosion phenomena. The combination of these natural and man-made factors has resulted in environmental and socio-economic damage.

PART THREE :

IMPACTS OF COASTAL EROSION AND ADAPTATION STRATEGIES IN THE COMMUNE OF KAFOUNTINE

The coastline of the Commune of Kafountine is being subjected to devastating assaults that will add to the current natural phenomena of erosion, degradation of mangroves and salinization of water and soil. In addition to natural factors linked to marine and subaerial agents, coastal erosion is accentuated by the acceleration of urbanization (residential, tourism and fishing) on the coast.

Faced with the scale of the ecological and socio-economic impacts of coastal erosion, local populations, decision-makers and investors are taking action to find solutions.

CHAPTER 1: THE ENVIRONMENTAL AND SOCIO-ECONOMIC IMPACTS OF COASTAL EROSION

The coastal zone of the Commune of Kafountine contains an ecological and biological heritage. It has become a place for human settlement and socio-economic activities that are essential to the area's development. With all this socio-economic development, man faces a major challenge. The sea's encroachment on the continent can have enormous repercussions on people's lives, plant and animal resources, and even the local economy. It is these environmental and socio-economic consequences that we will assess in this study.

1. Environmental impacts of coastal erosion

The consequences of coastal erosion are often catastrophic and vary spatially. Over the entire coastline of the Commune of Kafountine, they have distorted the geographical landscape. In environmental terms, coastal erosion is reflected in the mobility of the coastline, the destruction of coastal vegetation and the salinization of land and groundwater.

1.1. Coastline kinematics

Kinematics is the study of the dynamics of forms, independently of the forces that create them (Pennober, 1999; Akadje, 2010; Faye 2010). It translates into the erosion, accumulation or stability of forms.

Kinematic analysis requires first and foremost the choice of a coastline indicator. For this purpose, we have chosen the vegetation limit visible in satellite images as the coastline indicator. In the context of this study, the kinematics of the coastline is assessed by mapping the evolution of the coastline using Landsat satellite images acquired in 1979, 1986, 2000 and 2015, identifying areas of erosion and progradation, and calculating surface balances to estimate the areas lost or accumulated.

1.1.1 Coastline evolution and surface balances between 1979 and 2015

Due to erosive marine and anthropogenic factors, the coastline shifts over time. In its dynamics, the beach is either eroded, stable or nourished and changes morphology. We have subdivided the Commune's coastal zone into two sectors (the sector from Abéné to the bird peninsula and the sector from Ebédou to Diogué) to facilitate cartographic analysis.

❖ ***Period 1979-1986 :***

- From the Abéné coastal area to the bird peninsula :

The map showing the evolution of the coastline between 1979 and 1986 shows the progressive dynamics of the accumulating areas (map 5). To make the map easier to read, we have divided the area from Abéné to Presqu'île aux oiseaux into three zones.

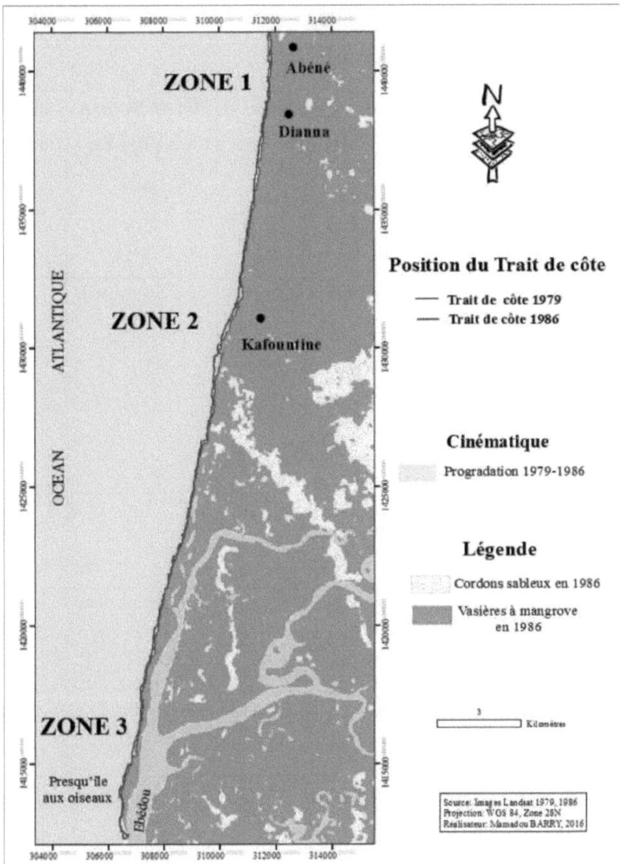

Map 5: Coastline evolution and surface area balance in the coastal sector from Abéné to the bird peninsula between 1979 and 1986.

Measuring differences between vegetation lines extracted from MSS (1979) and TM (1986) images shows a trend characterized by generalized progradation (Table 3).

Table 3: Statistics on shoreline dynamics from Abéné to Presqu'île aux oiseaux between 1979 and 1986

Zone	Location	Evolution average	Average growth rate	Growth rate Average for zone	Accumulated or eroded surfaces
Zone 1	Abéné	+471,41 m	+67.34 m/year	+60.84 m/year	+42.09 ha
	Dianna	+331,53 m +581,58 m +319,08 m	+47.36 m/year +83.08 m/year +45.58m/year		
Zone 2	Kafountine	+206,48 m +403,07 m	+29.50 m/year +57.58 m/year	+43.54 m/year	+26.74 ha
Zone 3	Presqu'île aux oiseaux	+272,81 m +139,21 m +265,02 m +164 ,32 m	+38.97 m/year +19.89 m/year +37.86 m/year +23.47 m/year	+30.05 m/year	+ 17.08 ha

Zone 1 recorded an advance of 425.88 m, i.e. an average progression rate of +60.84 m/year and an increase in surface area of 42.09 ha.

The average progression of zone 2 is +43.54 m/year, i.e. an extension of +304.78 m in linear distance and a surface area of 26.74 ha.

Zone 3 grew by 210.35 m, i.e. an average rate of +30.05/year and a surface gain of 17.08 ha.

The total surface area accumulated is estimated at 85.91 ha.

- From the coastal area of Ebédou to Diogué

Generally speaking, this sector is characterized by widespread accumulations (map 6). However, rates vary from zone to zone.

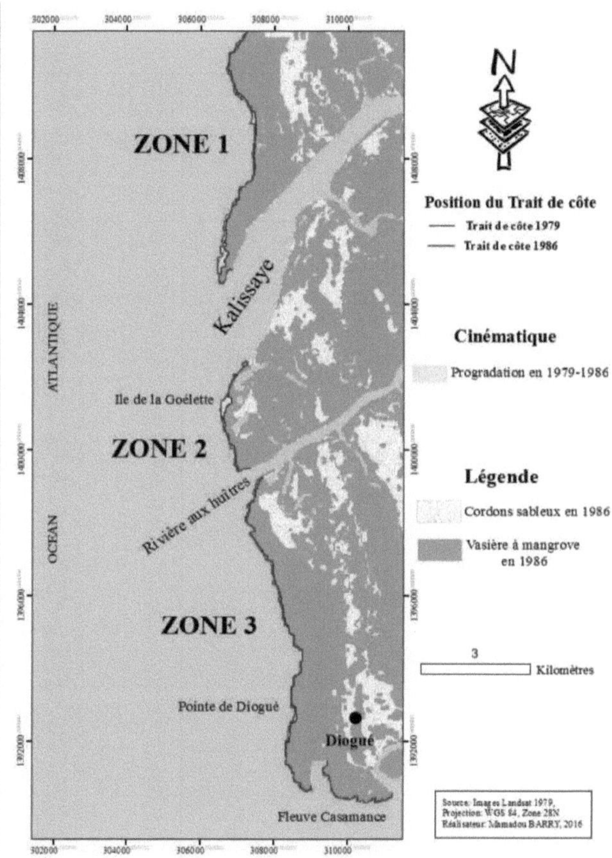

Map 6: Coastline evolution and surface balance in the coastal sector from Ebédou to Diogué between 1979 and 1986.

The evolution of the vegetation limit from Ebédou to Diogué shows a progressive dynamic of prograding zones concentrated in the free end of the sandy strips (table 4).

Table 4 Statistics of coastline dynamics from Ebédou to Diogué between 1979 and 1986

Zone	Location	Average trend	Average growth rate	Growth rate Average for zone	Accumulated or eroded surfaces
Zone 1	North of the mouth of Kalissaye	+265,82 m	+37.97 m/year	+41.93 m/year	+17.13 ha
	The Kalissaye spire	+321,23 m	+45.89 m/year		
Zone 2	South of the mouth of Kalissaye	+94.52 m + 55.61 m +264.41 m	+13.50 m/year +7.94 m/year +37.77 m/year	+19.37 m/year	+108.67 ha
	North of the mouth of the Rivière aux Huîtres	+74,56 m	+10.65 m/year		
	South of the mouth of the Rivière aux Huîtres	+188,83 m	+26.98 m/year		
Zone 3	Pointe de Diogué	+136,01 m +110,49 m	+19.43 m/year +15.78 m/year	+17.61 m/year	3.74 ha

Zone 1 recorded an average rate of +41.93 m/year m/year, corresponding to an extension of 293.51 m in linear distance and a surface gain of 17.13 ha.

Zone 2 advanced at an average rate of +19.37 m/year, with a gain of 108.67 ha.

Zone 3 grew at an average rate of +17.61 m/year, i.e. a surface area increase of 3.74 ha.

The total area accumulated in all zones is 129.54 ha.

1986-2000 :

- From the Abéné coastal area to the bird peninsula

The coastline evolution map between 1986 and 2000 is characterized by widespread accumulation and retreat of the coastline (map 7).

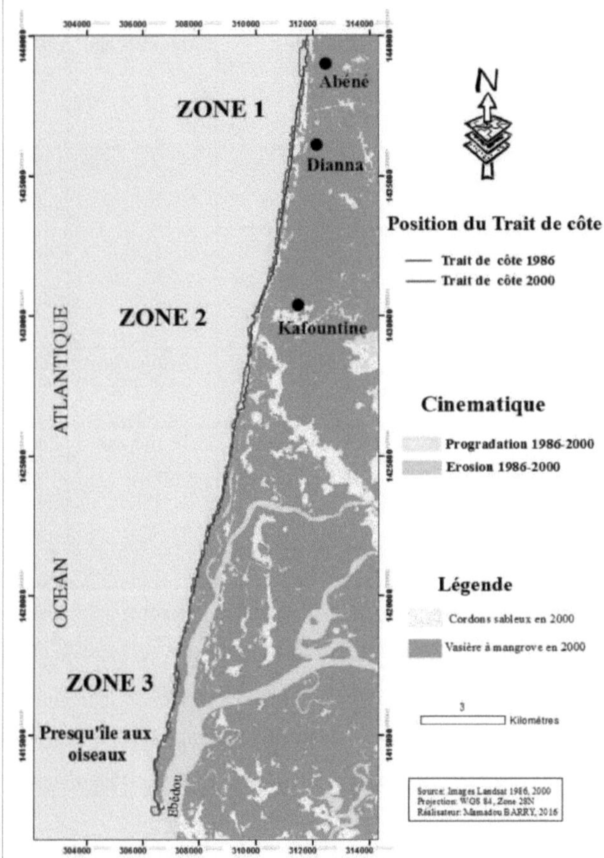

Map 7: Coastline evolution and surface area balance in the coastal sector from Abéné to the bird peninsula between 1986 and 2000.

A comparison of the vegetation boundary between 1986 and 2000 reveals a contrasting evolution, with a predominance of advancing and retreating phases (table 5).

<u>Table 5</u> Statistics on coastline dynamics from Abéné to Presqu'île aux oiseaux between 1986 and 2000

Zone	Location	Evolution average	Evolution rate Medium	Evolution rate Zone average	Accumulated or eroded surfaces
Zone 1	Abéné	+418,09 m	+29.86 m/year	+15.82 m/year	+58 ha
	Dianna	+123,83 m +172,26 m 178,12 m	+8.45 m/year +12.30 m/year +12.72 m/year		
Zone 2	Kafountine	+194,07 m +333,99 m	+13.86 m/year +23.86 m/year	+18.86 m/year	+33 ha
Zone 3	Presqu'île aux oiseaux	-72.27 m +112.95 m +79.82 m +89.40 m	-5.16 m/year +8.07 m/year +5.70 m/year +6.38 m/year	-5.16 m/year + 6.69 m/year	-2.62 ha + 14 ha

Zone 1 recorded an average rate of +15.82 m/year, representing an extension of 221.48 m and an increase in surface area of 58 ha. Zone 2 is also characterized by an increase of +264.04 m in linear distance, i.e. an average evolution rate of +18.86 m/year and a gain of 33 ha.

Zone 3 progressed at an average rate of +6.69 m/year, representing an increase in surface area of 14 ha. This resulted in erosion of 72.24 m, i.e. an average recession rate of -5.16 m/year, and a loss of 2.62 ha in surface area.

Accumulated surfaces are estimated at 105 ha and eroded surfaces at 2.62 ha.

<u>- From the coastal area of Ebédou to Diogué</u>

Generally speaking, this sector is characterized by widespread accumulation and retreat of the coastline (map 8). However, rates vary from one area to another.

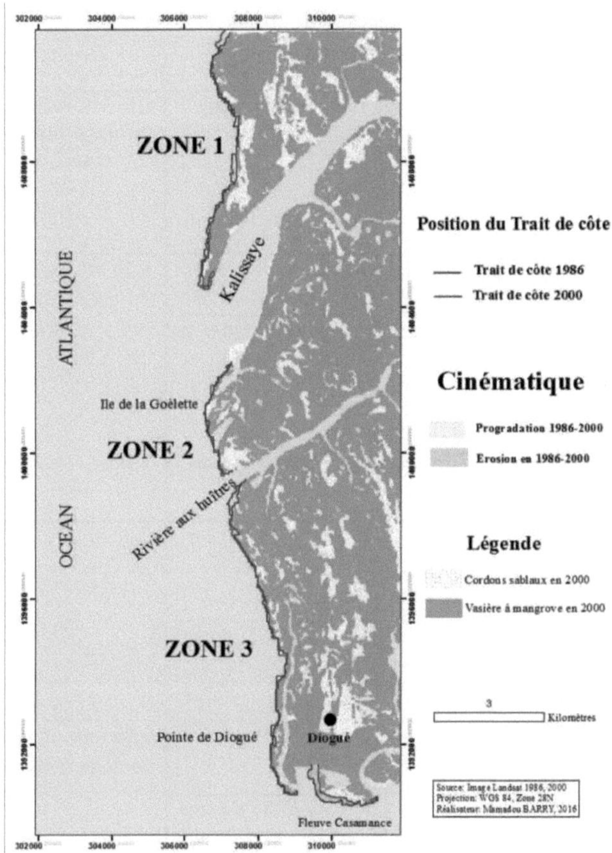

Map 8: Coastline evolution and surface balance of the coastal sector from Ebédou to Diogué between 1986 and 2000

The evolution of the vegetation boundary from Ebédou to Diogué shows a predominance of progradation concentrated in the free end of the sandy strips and lateral erosion (table 6).

Table 6: Statistics on coastline dynamics from Ebédou to Diogué between 2000 and 2005 2015

Zone	Location	Average trend	Average growth rate	Growth rate Average for zone	Accumulated or eroded surfaces
Zone 1	North of Ebedou	+ 122,96 m	+8.78 m/year	+8.76 m/year	+21.98 ha
	North of the mouth of Kalissaye	+152,17 m	+10.87 m/year		
	The Kalissaye spire	+92,64 m	+6.62 m/year		
Zone 2	South of the mouth of Kalissaye	+72,30 m	+5.19 m/year	+9.55 m/year	+16.73 ha
	South of the mouth of the Rivière aux Huîtres	+194.75 m +76.41 m	+13.91 m/year		
Zone 3	Pointe de Diogué	+54,46 m -59,51 m	+3.89 m/year -4.25m/year	+3.89 m/year -10.50 m/year	+2.01 ha -33.52 ha
	Southwest of Diogué	-333,84 m -47,80 m	-23.85 m/year -3.41 m/year		

Zone 1 recorded a growth rate of +8.76 m/year, representing an extension of 122.44 m in linear distance and an increase in surface area of 21.98 ha.

Zone 2 advanced at an average rate of +9.55 m/year, with a gain of +16.73 ha.

Zone 3 recorded an erosion rate of -10.50 m/year, i.e. an area loss of 33.52 ha, and an average progression rate of +3.89 m/year, i.e. an area increase of 2.01 ha.

Accumulated areas in all zones total 40.62 ha, while those lost are estimated at 33.52 ha.

Period 2000-2015 :

- From the Abéné coastal area to the bird peninsula

Depending on the area, this period may correspond either to a retreat of the coastline due to coastal erosion, or to an advance of the coastline due to sand accumulation (map 9).

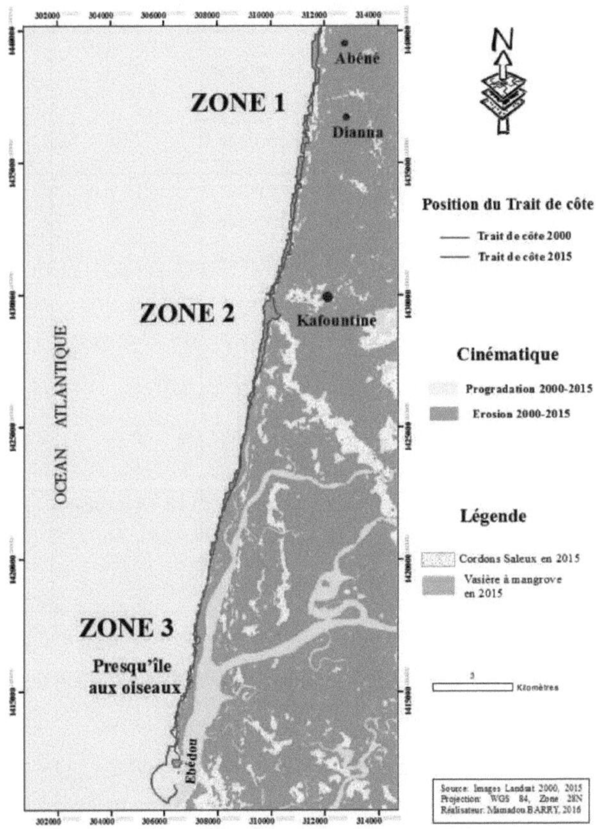

Map 9: Coastline evolution and surface area balance for the coastal sector from Abéné to the bird peninsula between 2000 and 2015.

The evolution of the coastline from the Abéné sector to the bird peninsula over the last fifteen years highlights a contrasting dynamic marked by a phase of retreat and a phase of advance (table 7).

<u>**Table 7**</u>: Statistics on coastline dynamics from Abéné to the bird peninsula between 2000 and 2015

<table>
<tr><th>Zone</th><th>Location</th><th>Average trend</th><th>Evolution rate Medium</th><th>Growth rate Average for zone</th><th>Accumulated or eroded surfaces</th></tr>
<tr><td rowspan="2">Zone 1</td><td>Abéné</td><td>-314,56 m</td><td>-20.98 m/year</td><td rowspan="2">-13.29 m/year
+10.62 m/year</td><td rowspan="2">-52.98 ha
+10.25 ha</td></tr>
<tr><td>Dianna</td><td>+159,32 m
-131,53 m
-151,81 m</td><td>+10.62 m/year
-8.77 m/year
-10.12 m/year</td></tr>
<tr><td>Zone 2</td><td>Kafountine</td><td>+35,50 m
-172,59 m
-253,65 m
-406,70 m
-705,09 m
-131,17 m
-255,06 m
-109,44 m</td><td>+2.37 m/year
-11.51 m/year
-16.91 m/year
-27.11 m/year
-47.01 m/year
-8.74 m/year
-17 m/year
-7.30 m/year</td><td>+2.37 m/year
-19.37 m/year</td><td>+0.95 ha
-146.48 ha</td></tr>
<tr><td>Zone 3</td><td>Presqu'île aux oiseaux</td><td>+440,39 m
+93,79 m
+59,16 m</td><td>+29.36 m/year
+6.25 m/year
+3.94 m/year</td><td>+13.18 m/year</td><td>+ 129.81 ha</td></tr>
</table>

The average rate of advance for zone 1 is +10.62 m/year, corresponding to a distance of 159.32 m and an area of 10.25 ha. It has also experienced erosion, with an average rate of retreat of -13.25 m/year, corresponding to a surface loss of -52.98 ha.

Zone 2 advanced at an average rate of 35.50 m, corresponding to a rate of +2.37 m/year and a surface gain of 0.95 ha. It was characterized by an average retreat of 209.53 m, corresponding to a retreat rate of -19.37 m/year and a loss of -146.48 ha.

Zone 3 upgraded by an average of 197.78 m in linear distance, i.e. an average rate of +13.18 m/year and a gain of 129.81 ha.

Total eroded area is estimated at 199.46 ha and accumulated area at 141.01 ha.

<u>- coastal area from Ebédou to Diogué</u>

On the beaches from Ebédou to Diogué, the same evolutionary trend can be observed, albeit to different degrees. However, the evolution recorded is completely opposite to that of the other sector (Map 10).

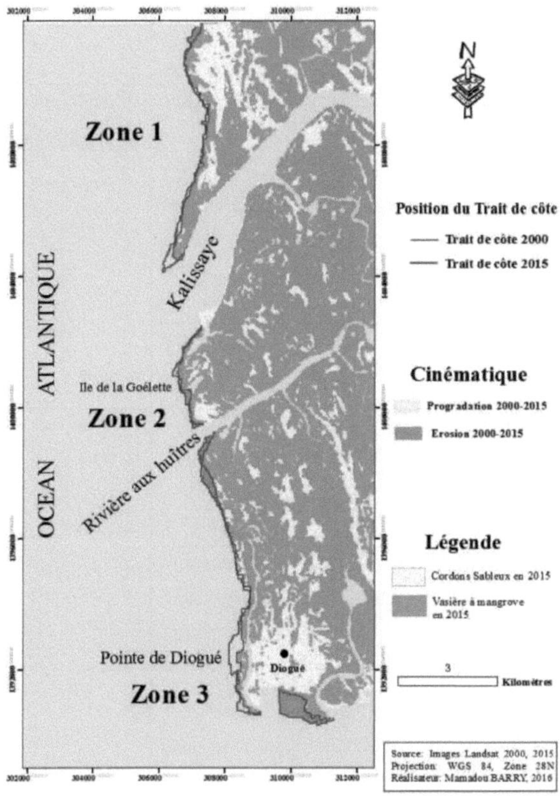

Map 10: Coastline evolution and surface balance in the coastal sector from Ebédou to Diogué between 2000 and 2015

Measuring the deviations between vegetation lines extracted from ETM+ (2000) and OLI (2015) images shows a regressive predominance despite progressive trends (Table 8).

Table 8: Statistics on coastline dynamics from Ebédou to Diogué between 2000 and present. 2015

Zone	Location	Average trend	Evolution rate Medium	Growth rate Average for zone	Accumulated or eroded surfaces
Zone 1	North of Ebédou	-248,37 m	-16.56 m/year	-16.56 m/year +11.37 m/year	-12.67 ha +23.21 ha
	North of the mouth of Kalissaye	+143,85 m	+9.59 m/year		
	The Kalissaye spire	+197,25 m	+13.15 m/year		
Zone 2	South of the mouth of Kalissaye	+64,34 m + 130,10 m	4.29 m/year +8.67 m/year	+5.24 m/year -10.33 m/year	+8.97 ha -44.51 ha
	North of the mouth of the Rivière aux Huîtres	+41,72 m -123,04 m -71,86 m	+2.78 m/year -8.20 m/year -4.79 m/year		
	South of the mouth of the Rivière aux Huîtres	-183,72 m -241,01 m	-12.25 m/year -16.07 m/year		
Zone 3	Pointe de Diogué	+323,02 m +114,67 m	+21.53 m/year +7.64 m/year	+14.59 m/year -24.20 m/year	+49.46 ha -65.87 ha
	southwest of Diogué	-363,07 m	-24.20 m/year		

Zone 1 is characterized by an average regression of -248.37 m, corresponding to an average rate of retreat of -12.64 ha and a surface loss of -12.67 ha. It advanced at an average rate of 170.55 m in linear distance, an average rate of progression of 11.37 m/year and a surface gain of 23.21 ha.

Zone 2 recorded an average advance of 70.60 m, i.e. a rate of 5.24 m/year, corresponding to a gain of 8.97 ha, and an average retreat of -10.33 m/year, with a loss of 44.57 ha.

Zone 3 is characterized by a progradation rate of 14.59 m/year, corresponding to an extension of 218.85 m in linear distance and an increase in surface area of 49.46 ha. It also recorded an average recession rate of -24.20 m/year, corresponding to a regression of 363.07 m in length and a surface loss of 65.87 ha.

Surface losses are estimated at 123.05 ha, while accumulated surfaces represent a total gain of 81.64 ha.

I.1.2. Summary of surface area balances between 1979 and 2015

The coastline of Kafountine Commune has experienced a certain dynamic in terms of surface area over the last 36 years. The surface balance of the coastline has undergone two phases: an accumulation phase between 1979-1986 and 1986-2000, and an erosion phase between 2000 and 2015 (figure 5).

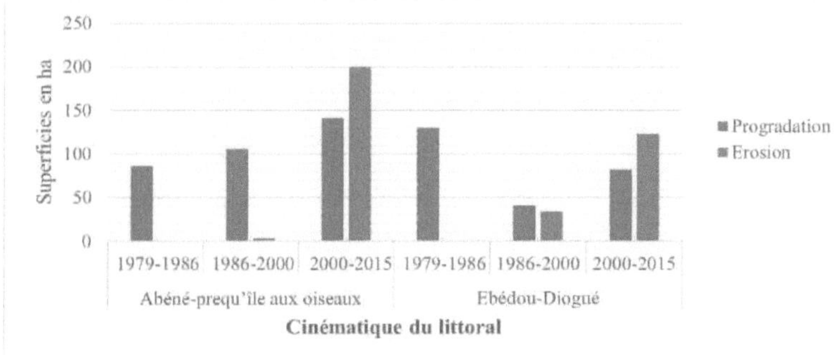

Figure 5: Evolution of the surface area balance of the coastline of the Commune of Kafountine between the periods 1979, 1986, 2000 and 2015

In the Abéné to bird peninsula sector, progradation increased from 86.91 ha between 1979 and 1986 to 105 ha between 1986 and 2000, representing an increase of 18.09 ha or 20.81%. Between 1986-2000 and 2000-2015, the coastline increased from 105 ha to 141.01 ha, representing a surface gain of 36.01 ha. In the same sector, erosion rose from 2.62 ha in 1986-2000 to 199.46 ha in 2000-2015, representing a loss of 196.42 ha. This increase in coastal erosion in the Abéné to bird peninsula sector could be linked to the anthropic pressure exerted on the 72 coastline. The blossoming of tourist infrastructures, the construction of a fishing wharf at Kafountine in 2009 and the construction of related infrastructures (petrol stations, refrigeration plants) have led to an imbalance in the coastal system, which in turn has resulted in sediment deficits on the beach.

Progradation in the Ebédou to Diogué sector fell from 129.54 ha in 1979-1986 to 40.64 ha in 1986-2000, a drop of 88.9 ha. This drop in surface area can be attributed to the aggressiveness of hydrodynamic agents, which can cause sediment deficits. Between 1986-2000 and 2000-2015, the progradation trend increased from 40.64 ha to 81.64 ha, a gain of 41 ha. This increase in surface area is linked to the reforestation of coastal dunes and mangroves on mudflats. On the other hand, the area eroded rose from 33.52 ha between 1986 and 2000 to 123.05 ha between 1986 and 2000, i.e. a loss of 89.53 ha. This increase in eroded areas is attributable to Diogué's exposure to the mouth of the Casamance River, which is driven by hydrodynamic phenomena that erode the coastline. Anthropogenic factors, such as sea sand extraction and mangrove deforestation, are also responsible.

1.2. Degradation of vegetation

The impact of erosion on vegetation manifests itself as an ecological crisis. Generally speaking, an ecological crisis occurs when the biophysical environment of an individual, a species or a population of species changes in a way that is unfavorable to its survival (Diakhaté, 2008 in Thior 2013).

- Land use in the Commune of Kafountine between 1979 and 2015

In the Commune of Kafountine, mangroves and other vegetation types have evolved over the last 36 years. The results of the various treatments have enabled us to produce multi-date maps (Map 11) showing land use, and in particular vegetation, for each year in our series (1979, 1986, 2000 and 2015).

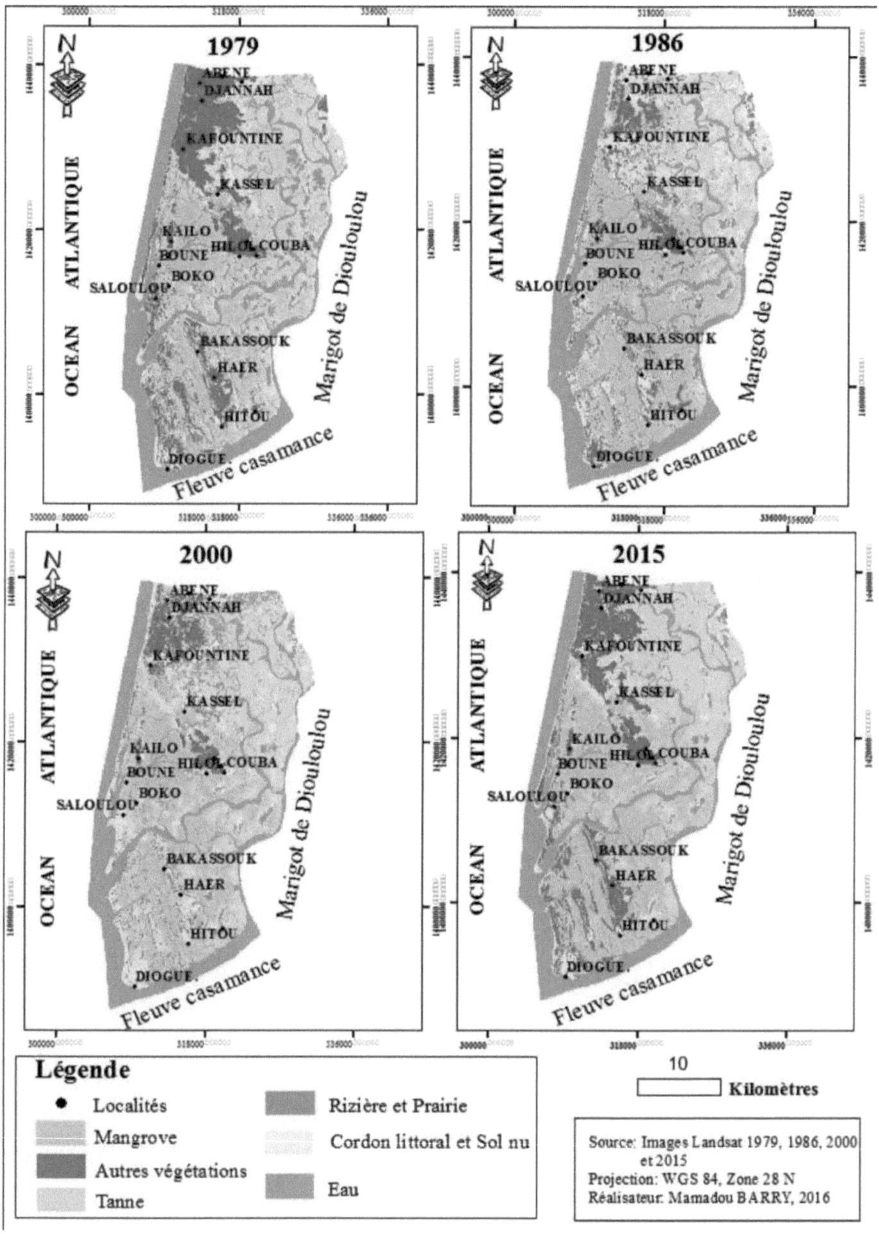

Map 11: Land use in the Commune of Kafountine in 1979, 1986, 2000 and 2015.

Land-use maps for the Commune of Kafountine in 1979, 1986, 2000 and 2015 show, in terms of surface area, a great dynamic (Figure 6) of the different land-use classes, particularly mangroves, tannes and other vegetation types.

Figure 6: Change in area of different land-use classes in the

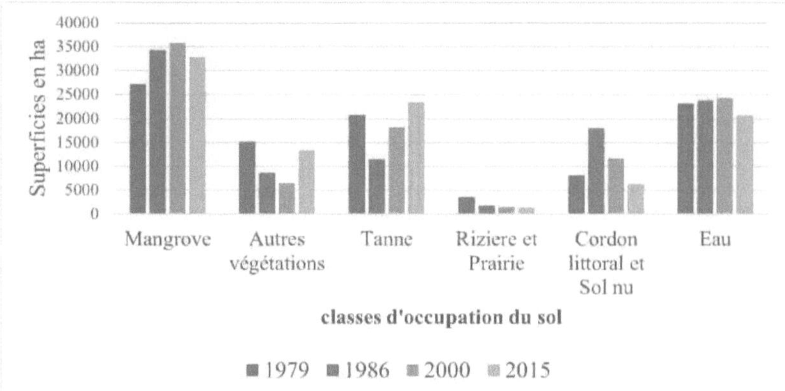

Commune of Kafountine in 1979, 1986, 2000 and 2015.

Over the last 36 years, the mangrove in the Commune of Kafountine has undergone two phases of evolution: an increase between 1979, 1986 and 2000, and a slight decrease between 2000 and 2015.

In 1979, the mangrove area was much smaller than in other years. This area was estimated at 27,166 ha. This deterioration was the result of the drought of the 1970s and 1980s. This manifested itself in a drop in rainfall, with consequences for the mangrove.

However, in 1986, the mangrove regenerated. It increased from 27166 ha in 1979 to 34304 ha in 1986. During this period, the mangrove increased by 26.28%, i.e. an area of 7138 ha. This increase was accompanied by a reduction in the area of tannes, from 20690 ha in 1979 to 11563 ha in 1986, i.e. a decline of 44.11%. As in 2000, the mangrove regenerated. It covered an area of 3,4304 ha in 1986. This has increased by 4.34% to 35793 ha. This regeneration of the mangrove is linked to the return of rainfall, which creates favorable conditions for this ecosystem to develop properly. In addition, awareness of the importance of mangroves has been raised through restoration and conservation actions.

In 2015, the mangrove regressed very little. Over a 15-year period, it covered 35793 ha in 2000. By 2015, it had shrunk to 32747 ha, a regression of 8.51%, representing an area of 3046 ha.

This deterioration has been to the benefit of the tannes, which have seen an increase in surface area. In 2000, they occupied an area of 1,8249 ha, and in 2015, 2,3409 ha. This represents an increase of 28.28%, or 5,160 ha.

This reduction in mangrove area is linked to mangrove silting (Photo 5). Indeed, with the widening of river mouths and active sedimentation on the sea front. We are witnessing the removal of sandy particles from the banks, which are drained into the mangrove perimeters, mixing with the silt. Diogué Island, located at the mouth of the Casamance River, receives a lot of sandy particles from its watershed. This situation is at the origin of a regressive evolution of the mangrove.

Photo5: Silting of mangrove roots in Diogué (Barry, June 2016)

In addition to silting, salinization is an explanatory factor in mangrove dynamics. In the Casamance estuary, salinity is highly variable and seasonally determined. According to Leroy *et al* (1986), salinity in Casamance has led to the shrinkage of *Rhizophora* mangroves on the main inlets and mudflats, their almost total disappearance on the secondary inlets, and their replacement by *Avicennia* mangroves better adapted to excess salt.

In addition, the other vegetation types in the Commune of Kafountine are characterized by two phrases (Figure 6): a regression phrase between 1979, 1986 and 2000, and an increase between 2000 and 2015.

In 1979, other vegetation types were higher than in other years, with a surface area of 15158 ha. In contrast to the previous year, other vegetation types declined in 1986. Its surface area covered 8548 ha, a drop of 43.61%. During this period, they suffered a decline of 6,610 ha. As in 2000, other vegetation types also declined. They represented an area of 8548 ha in 1986. In 2000, they fell to 6414 ha, a regression of 24.96%, representing 2134 ha. This considerable decline in other vegetation is attributable to deforestation in upland areas and destructive wave action in coastal areas.

In 2015, the regeneration of other vegetation rose from 6414 ha in 2000 to 13371 ha in 2015, an increase of 108.47%. This extension of 6957 ha is the result of vegetation restoration and conservation actions.

In the coastal zone of the Commune of Kafountine, coastal erosion is manifested by the destruction of coastal vegetation (Photos 6 and 7). This vegetation, which defends the coast, is subject to wave action, causing the coastline to recede.

Photo 6: Destruction and significant mortality of the species *Casuarina equisetifolia Filao)* at Abéné (A) and Diogué (B) (Barry, March and June 2016)

Photo 7: Mortality of *Elaeis guineensis* (palms) and *Coco nucífera* in Diogué (Barry, June 2016)

Loss of coastal vegetation along the Kafountine coast has led to weakening of the dunes and

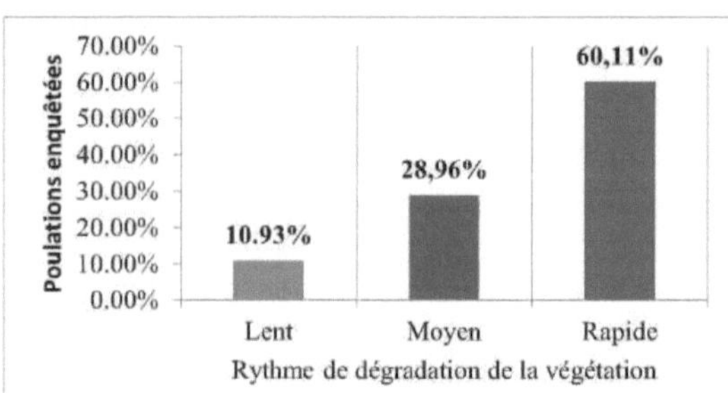

accelerated coastal erosion.

A large proportion of the population surveyed (60.11%) stated that coastal erosion is destroying vegetation at a rapid rate (figure 7). This ecological situation can result in a loss of biodiversity, which is often felt in the Commune's socio-economic activities.

Figure 7: Vegetation mortality rate in the Kafountine coastal zone.

(Source: Survey data, 2016)

- Spatial dynamics of mangrove and other vegetation between 1979 and 2015 To produce the change maps, we used five (5) land cover classes: mangrove, other vegetation types, tanne, water and sandy strips. Each of these classes was assigned the same code so that images from 1979 and 2015 could be cross-referenced by addition.

The statistical results of these cartographic processes have enabled us to determine the stable, appeared and disappeared areas of mangrove and other vegetation during the period 1979 and 2015 (maps 12 and 13).

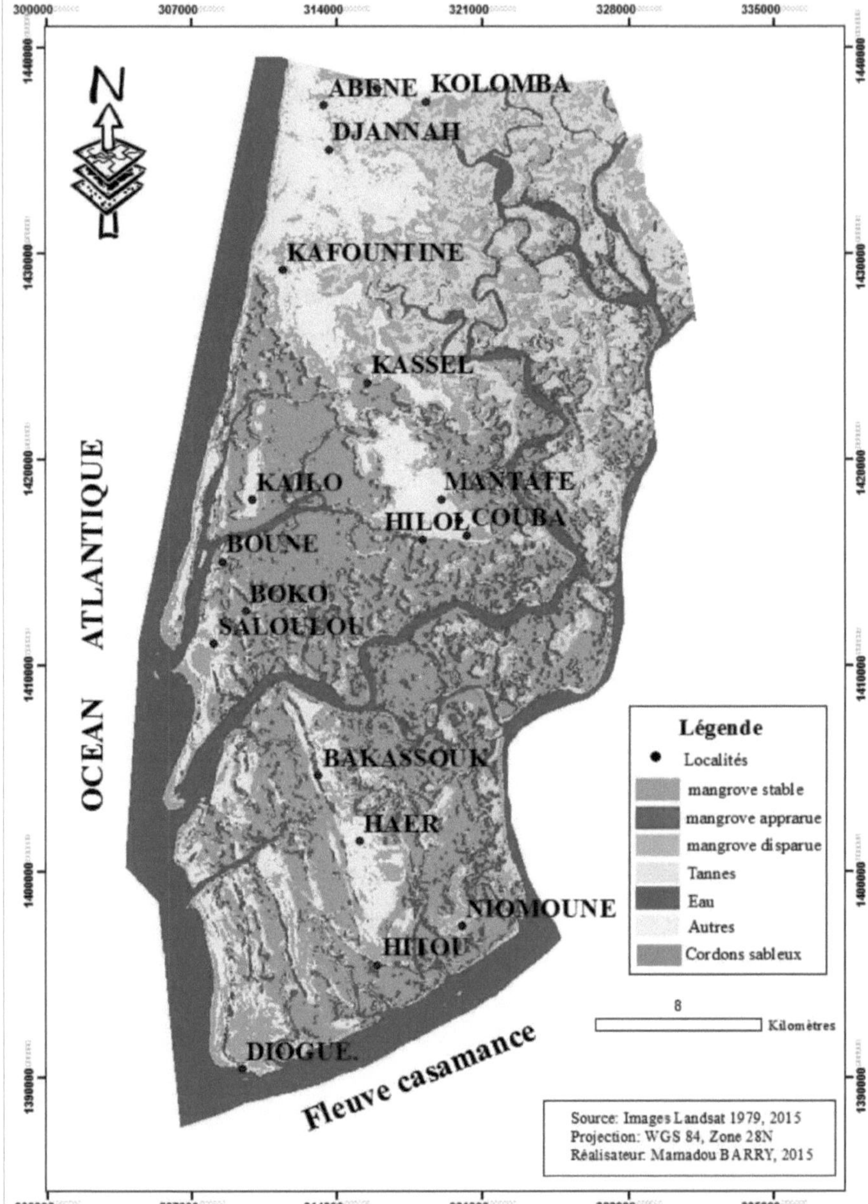

Map 12: Mangrove dynamics in the Commune of Kafountine between 1979 and 2015

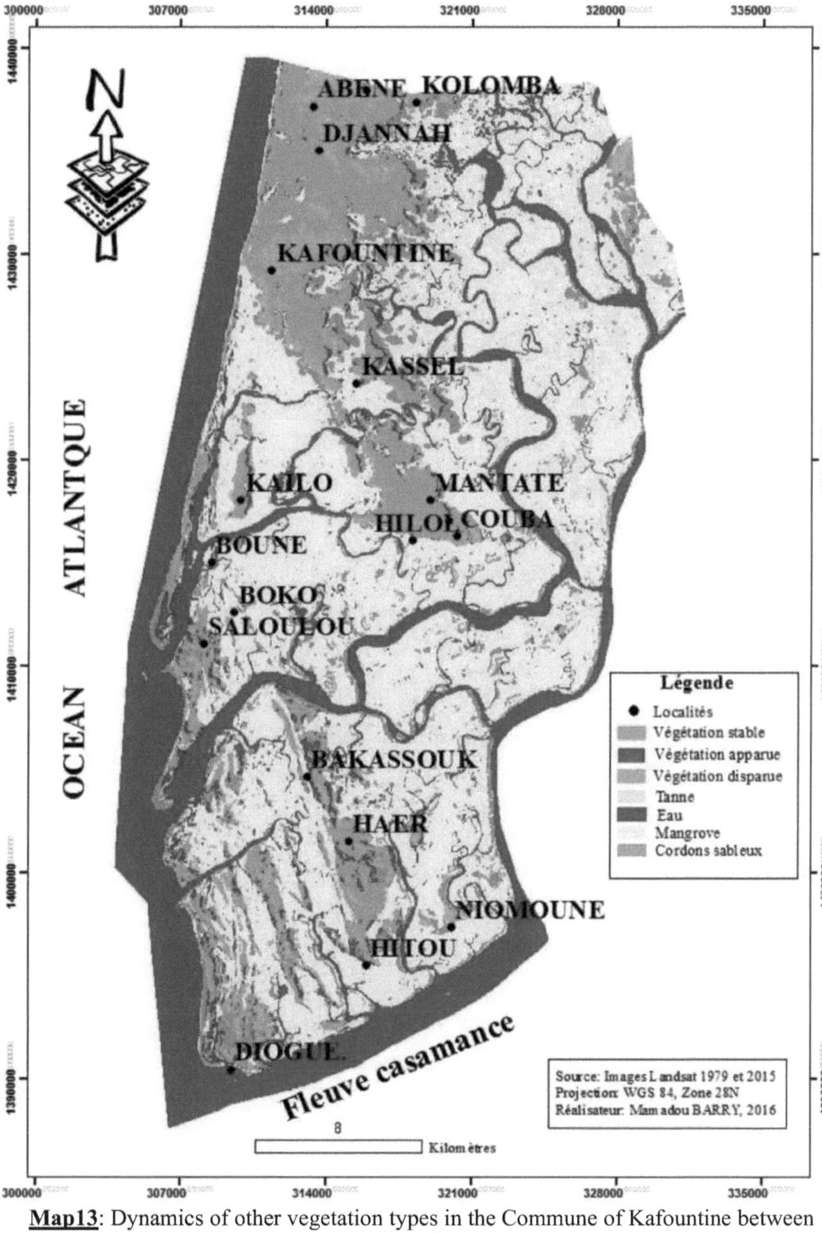

Map13: Dynamics of other vegetation types in the Commune of Kafountine between 1979 and 2015

The maps of changes in mangrove and other vegetation in the Commune of Kafountine in 1979 and 2015 show variable surface areas in terms of spatial dynamics. Thanks to these changes, each interval of stable, appeared and disappeared areas is precisely known (Figure 8).

Figure 8: Spatial evolution of change classes between 1979 and 2015.

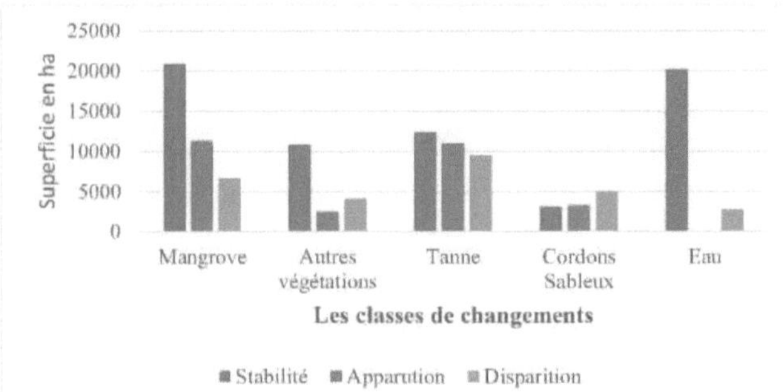

Over the last 36 years, the results of changes have shown three evolutionary phases for mangroves and other vegetation.

Between 1979 and 2015, stable mangrove areas covered 20750 ha, areas that had appeared covered 11243 ha and areas that had disappeared covered 4093 ha. The stability and appearance of the mangrove are the result of a return of rainfall favorable to mangrove regeneration, but also to reforestation actions and awareness of the ecological importance of the mangrove.

Other vegetation types remained stable at 10748 ha, appeared on 2246 ha and disappeared on 4093 ha. This low appearance of other vegetation types, alternated by a significant disappearance, is linked to the drought of the 1970s and 1980s, but also to anthropogenic factors such as deforestation.

1.3. Environmental impact of salinization of rice paddies

In the Casamance estuary, salinization is linked to the penetration of the salt tongue from the Casamance River, which is facilitated by low freshwater inflows.

Indeed, with the lengthening of the drought (Marius, 1986; Mahé, 2006) and the flat topography of the Casamance basin (Dacosta, 1989; Diop, 1990), the hydrographic network of the Casamance River is increasingly influenced by the tide (Montoroi, Flows, already low, have decreased and freshwater inflows have become insufficient to compensate for the upwelling of sea water (Salem, 1992 in Sarr, 2012).

In the Commune of Kafountine, flooded rice-growing plays an important role, and is practised in fluvial areas. In these areas, where tidal variations increase the frequency of salt in the soil, agriculture is currently struggling to develop. A number of hypotheses are put forward by local people to explain the causes of salinization:

- salinization is due to reduced rainfall;
- the proximity of the rice fields to the river;
- mangrove degradation;
- salt rises by capillary action.

The results of map processing (map 11) and (figure 6) of land use in 1979, 1986, 2000 and 2015 show that in 1979, rice paddies and grasslands occupied an area of 3609 ha. In contrast to 1979, rice paddies and grasslands declined in 1986. They covered 1,742 ha, down 51.73% to 1,867 ha.

In 2000, there was a slight decline in the area under rice paddies and grasslands, from 1,742 ha in 1986 to 1,452 ha in 2000, a drop of 16.65%.

Rice paddies and grasslands experienced the same regressive trends in 2015. Covering an area of 1,452 ha in 2000, they fell to 1,394 ha in 2015, a loss of 3.99% or 58 ha.

These losses are offset by the expansion of tannes. The latter have evolved from 18249 ha in 2000 to 23409 ha in 2015, an increase of 28.28%. This reduction in rice paddy area is linked to salinization. Rice paddies are sensitive to rainfall variations and high tides, which can lead to salinity in rice fields. This could result in the widening of the tannes to the benefit of the rice fields and a lack of productivity in the rice fields.

1.4. Impact of salinization on water tables

The dynamics of the water table are essential to the evolution of this environment. The shallow slopes of the lowlands encourage the penetration of sea water. Saltwater, with an average density of 1.025 kg/l (compared with 1.000 kg/l for freshwater), tends to penetrate inland under the action of the density gradient (Dreal, 2011). In mangrove areas, the seasonal amplitude of the water table is less than one meter in the tannes, and two meters in the terraces (Sarr, 2012). According to Mougenot (1966), this proximity, during the seven to eight months of the dry season, favors the concentration of salts in the upper parts of the water table, with saline upwelling to the soil surface resulting in a lack of desalination in the rainy season.

In the island communities of the Commune of Kafountine, people rely on water from wells. The problem of drinking water supply has become a reality in villages located close to the sea. Indeed, most water resources are affected by salinization due to the drying up of freshwater aquifers and the rise of the salt water table. The freshwater table dries out too quickly, and is gradually replaced by the saltwater wedge.

II. Socio-economic impacts of coastal erosion

As in other coastal regions of Senegal, the coastal zone of Kafountine Commune is the site of major economic activities.

Today, these various socio-economic activities, which play a key role in the Commune's economy, are threatened by the ever-increasing coastal erosion in the area. The consequences are highly visible in the Commune, and include the degradation of rice-growing activities, the destruction of tourist infrastructures and the deterioration in the quality of water intended for consumption.

11.1. Falling rice yields

Land degradation is one of the consequences of the advancing sea on rice-growing land. Land degradation has a negative impact on agriculture, which plays a key role in the economy.

In the Commune of Kafountine, alluvial zones have long been naturally favorable to rice-growing activities. Favorable factors included their relatively low-lying topography, the natural fertility of the soils concerned, relatively high and regular rainfall, and the presence of a long-standing rice-growing tradition among the local population. Today, the various droughts (Montoroi, 1991) and the upwelling of salt water into the Casamance river system have led to the salinization of the rice fields. As a result, local populations have seen their useful rice-growing areas shrink considerably over the years. In the Commune, island areas are the hardest hit by declining yields and abandonment. rice-growing land. Indeed, most of those interviewed said they had abandoned rice plots because of salinity and toxicity (figure 9).

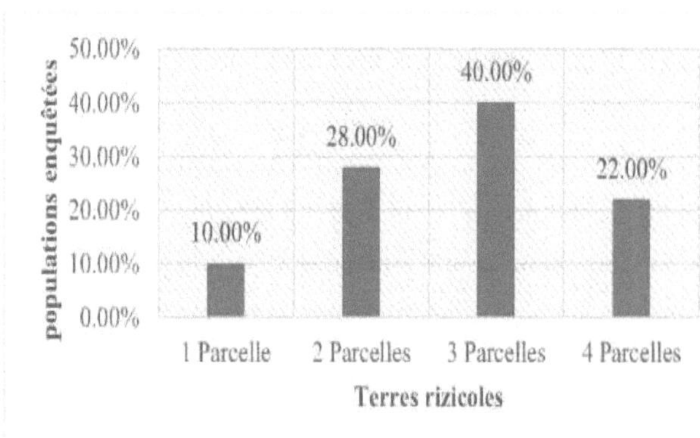

Figure 9: Loss of rice-growing land in the Commune of Kafountine

(Source: Survey data, 2016)

Household surveys showed that 40% of heads of household had lost at least three rice fields (figure 9). According to the local population, this phenomenon is the result of the advanced state of degradation of rice field soils, which have a salinity level that far exceeds the rice's tolerance level. An excess of water-soluble salts increases osmotic pressure, making it difficult for plants to absorb water. Even if the soil appears to have plenty of moisture, the plants wilt because the roots do not absorb enough water to replace that lost through evapotranspiration.

This often leads to malformation of the grains, and consequently a drop in rice yields (Figure 10), with enormous consequences for the livelihoods of the people who depend on this activity.

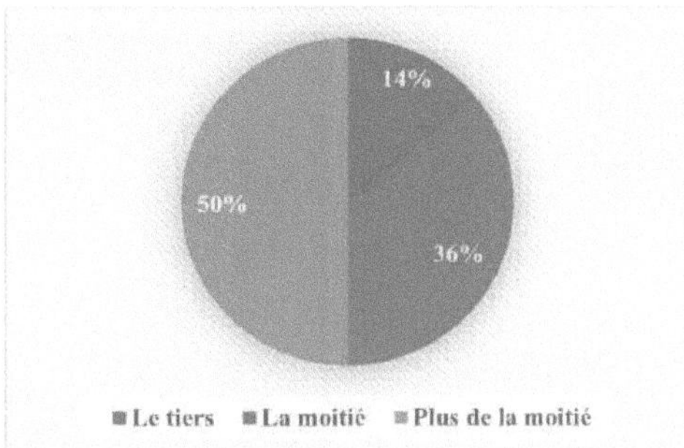

Figure 10: Decline in rice production in the Commune of Kafountine

(Source: Survey data, 2016)

This drop in yields and the loss of rice-growing areas have led to the islanders' withdrawal from rice-growing, and they are no longer able to ensure rice self-sufficiency.

This drop in yields has also resulted in young people, who are the main labor force for rice cultivation, leaving for the towns in search of better financial conditions to make up for the drop in income.

Today, with the decline in rice-growing activity, these migrations, once seasonal, are tending to become permanent. Those who leave no longer wish to return to work in the rice fields. This situation has led to a shortage of labor in the villages to cultivate and maintain the rice plots. The anti-salt dykes are less and less functional, which favors the intrusion of salt water into the rice fields.

11.2. Impact on tourism infrastructure

Tourism is an activity that generates jobs and income in the Commune of Kafountine. It plays an important role in the local economy, with many young people involved.

In fact, the majority of tourist facilities are built on the upper beach (map 3).

This tourism sector is currently suffering from marine erosion, which is threatening tourist infrastructures.

In Abéné, for example, the Atlantic Club camp is just a few meters from the water (Photo 8). The now-abandoned Kalissaye hotel is also affected by coastal erosion.

Photo 8: Atlantic Club Hotel a few meters from the shore threatened by coastal erosion (Barry, March 2016)

On the coastal fringe of Kafountine, the phenomenon is far more alarming. Coastal erosion has had a major impact on tourism. The Nature camp has been hit and partially destroyed by the waves (Photo 9).

Photo 9: La Nature camp destroyed by coastal erosion in Kafountine (Barry, March 2016)

In Diogué, which faces the mouth of the Casamance River, buildings are being undermined by waves that threaten them from day to day. Field surveys carried out in this locality have shown that some houses have been abandoned because of the advancing sea (Photo 10), and some buildings, such as the elementary school, which more than 20 years ago was located 3 km from the sea, is now 40 meters from the coast.

Photo 10: Houses abandoned by fishermen in Diogué (Barry, January 2017)

11.3. Difficult access to drinking water

Water is a development factor and an indispensable natural resource that guarantees the life and well-being of populations. On the islands of the Commune of Kafountine, the main aquifers are closely linked to watercourses. Their recharge is governed mainly by rainfall and infiltration.

Groundwater in the mainland (Kafountine, Abéné, Albadar and Colomba) is of good quality and suitable for all uses. Aquifers are used extensively for citrus cultivation, as well as for market gardening. Wells 6 to 10 meters deep have been built for fruit growing and market gardening (especially in Albadar, Dianna and Colomba). For drinking water, two boreholes, one at Kafountine and the other at Abéné, each with a capacity of 150 m^3 supply drinking water to the two localities.

In island areas, on the other hand, groundwater is saline due to saltwater intrusion and is of poor quality (Figure 11).

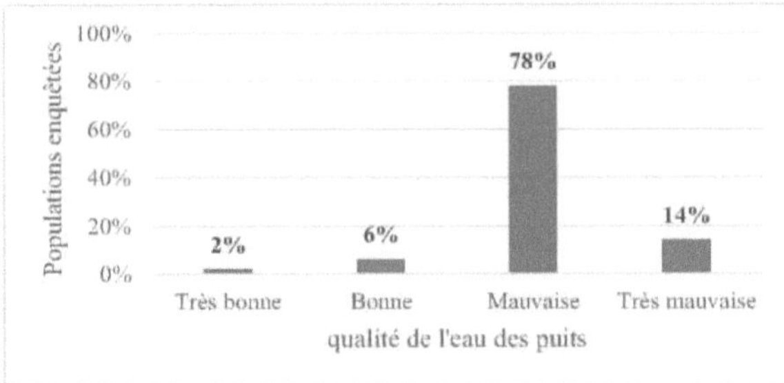

Figure 11: Well water quality on the islands (Source: Survey data, 2016).

Many islands have aquifers, but since these are freshwater lenses resting on top of a saltwater body, they are vulnerable to intrusions (Pignal, 2005). As a result, the entire freshwater potential, both at the surface and at depth, is seriously affected by salt. On most of the islands (Diogué, Mantate, Hilol, Saloulou and Niomoune), the water table is shallow. As a result, the quality and quantity of the well water that supplies the population fluctuates with the seasons.

During the dry season, from October to June, access to drinking water is a real problem. The water table drops and well water becomes brackish, sometimes muddy and unfit for consumption. This is why some islanders seek water in neighboring communities. Rainwater harvesting through tanks installed on the roofs of houses represents an important source of drinking water for the low-lying islands of Diogué, Boko and Kailo.

Conclusion

Analysis of satellite images from Landsat 1979, 1986, 2000 and 2015 has highlighted the evolution of the coastline, land use and vegetation dynamics.

This study shows that the evolution of the coastline varies from period to period, but also under the effect of human actions. From 1979, 1986 and 2000, we note a gradual dynamic resulting in accumulation. Between 2000 and 2015, on the other hand, the coastline tended to retreat as a result of construction work on the fishing pier and tourism. This mapping has also shown the dynamic evolution of mangroves, other types of vegetation, rice paddies and tannes. This dynamic is influenced by both natural and anthropogenic factors. It has led to a disruption of activities and water quality. This situation has led people to develop structural and non-structural strategies.

CHAPTER 2: STRATEGIES FOR ADAPTING TO COASTAL EROSION IN THE COMMUNE OF KAFOUNTINE

Coastal erosion in the Commune of Kafountine can be explained by a combination of natural and anthropogenic factors. Its environmental consequences have had an impact on socio-economic activities. To deal with this situation, the local population, local authorities and partners have implemented structural and non-structural methods to combat coastal erosion.

I. Structural methods

These options involve the construction of hard protective structures based on cement, reinforced concrete and more or less rigid natural rock defences.

In the Commune of Kafountine, most of the protective structures built to combat coastal erosion are the result of individual action, mainly by the tourism sector, to protect threatened infrastructure. However, these personal, ad hoc initiatives may solve an immediate problem, but they are by no means a lasting solution.

- Rigid structures

They harden the shoreline and are the most widely used in the area. They take the form of spurs and walls.

A groin is a protective structure perpendicular to the coast, designed to intercept sediments transported by longshore drift in order to reconstitute or modify the beach. During the dry season, when north-westerly swells and north-south lateral currents dominate, the groin captures some of the sediment transported, fattening the beach upstream of the obstacle. The main disadvantage of this type of protective structure is that it causes erosion downstream. However, during the winter months, when south-westerly swells dominate with a south-north drift, the opposite phenomenon occurs.

Protective walls are vertical structures that stabilize the shoreline and protect land and infrastructure from high levels of intensity. They are costly structures, requiring constant and irregular maintenance to extend their lifespan. A case in point is the owner of the "La Nature" hotel, Mr. Jean Louis Rambert, who built a groin in 2003 to protect his infrastructure against the devastating actions of the waves, then replaced it in 2009 with another vertical concrete protection wall. However, these costs only served to accelerate coastal erosion and the destruction of his hotel (Photo11).

Photoll: Destruction of La Nature hotel by waves at Kafountine: destruction of herringbone installations (A) and protective wall (B)

These structural countermeasures failed to stand up to the devastating action of the waves. These relatively rigid protective structures are becoming increasingly rare in the Commune due to the high costs involved. They are gradually being replaced by less costly and more flexible traditional methods.

Protective structures built by individual hoteliers often have an impact on the coastal system. Such protective structures should therefore be subject to an impact study.

The purpose of such a study is to provide detailed, rational answers on the sedimentary behavior of the coastline. It is only once this impact study has been carried out that the work should be launched, in compliance with the instructions for carrying out this mission and bypassing the harmful tendencies that such work could raise. Better coastal management requires knowledge of the physical environment and its reaction to natural agents, in order to avoid costly errors in planning and siting.

II. Non-structural methods

In the Commune of Kafountine, the non-structural or flexible methods focus on the vegetal option. It is traditionally used to retain sediments more effectively and

to combat coastal erosion. It's a protection and adaptation strategy that has the advantage of being within everyone's reach, without generating major negative effects.

II.1 . Fixation of sand bars

Biological fixation or reforestation of the dune cordon is a sand-fixing operation that is essential for beach protection. The coastline of the Commune of Kafountine is characterized by loose rock. This type of sediment, which forms the bulk of the landscape, explains the vulnerability of the coast, which cannot withstand the repeated assaults of the waves. In response to this situation, 2,6509

Niaoulis (*Melaleuca leucadendron*) seedlings were planted (Photo 12) on 5,000 m of dune cordon between Abéné, Diannah and Kafountine by the Association des Zones Humides du Sénégal (AZOHS) with the collaboration of the PRCM and the support of 23,000 people living in the Commune (women processors, fishermen, fish smokers, etc.).

Photo 12: Fixation of coastal dunes in Kafountine by planting *Melaleuca leucadendron* (Source: http://www.prcmarine.org)

Also, 54,068 filaos (*casuarina equisetifolia*) were planted over an area of 86.507 ha in the village of Diogué between 2014 and 2015 by the NGO Justice et Développement under the technical supervision of the Division Régionale de l'Environnement et des Etablissements Classés de Ziguinchor and the Service des Eaux et Forêts, but also with the support of local people (Photo 13).

Photo 13: Fixing sand dunes with a *casuarina equisetifolia* plantation in Diogué (Barry, June 2016)

In addition, individual hotel owners are increasingly taking part in reforestation initiatives. Baobab, Caserène, Casamar, Dianna Bolong and Atlantic Club, for example, have reforested certain plant species (Filaos, coconut palms, niaoulis, etc.) within their infrastructures to prevent the advance of sand dunes and counteract the action of sea winds. Fixing sandy strips using biological techniques on the coastal fringe of the Commune of Kafountine has stabilized loose material and acted as a windbreak to protect infrastructures and market garden crops. This strategy has also mitigated the damage caused by coastal erosion.

However, the plants used to reforest the dunes are not very resistant, as they can be damaged during periods of very high seas. In addition, there is little research and experimentation in coastal environments on the effectiveness of vegetation as a protection against erosion (ZIP, 2008).

In addition, the reinforcement of dune reforestation in the Commune of Kafountine contributes to the sustainable management of the coast. The Kafountine coast is a loose substrate that is highly sensitive to wind, so a solution must be found to fix the dunes and prevent them from moving inland. Our choice is *casuarina equisetifolia* (Filao).

Casuarina equisetifolia, from its roots to its leaves, has three ways of combating erosion:

- Filao develops a lateral root system that is both long and several metres deep. The importance of these roots is to trap sediment and protect it from wind and water erosion;
- the Filao has a vertical development that can reach a certain height. It serves as a curtain that can attenuate wind energy;
- good leaf development and heavy leaf fall provide ground cover. Once on the ground, these leaves form litter, preventing sediment from being dragged along by receding waves or carried away by wind action.

In order to manage this monospecific plantation (*casuarina equisetifolia)* properly, emphasis should be placed on diversifying the species used to fix the dunes, by giving more and more room to endogenous species and, *de facto,* reducing the proportion of exotic species. In this way, local species such as ipomoea pes-caprae, cocos nucifera and Borassus can coexist with Filaos.

II.2 . Mangrove reforestation

Mangrove restoration is a defence against coastal erosion. It encourages the accumulation and fixation of marine sediments to reduce the impact of coastal erosion. By sequestering atmospheric carbon, mangroves help mitigate global warming and counteract sea-level rise (Sow, 2010)[5] .

In the Commune, mangroves are planted both on the islands and on the mainland. It plays a particularly important role in protecting the coastline from the effects of tides and storms, as well as stabilizing the soil. Its halophilic nature enables it to resist hypersalinization. Between 2012 and 2015, the local population, with technical support from the Aire Marine Protégée d'Abéné and financial backing from WWF/Germany and WIA, reforested more than 200,000 species of *Rhizophora* and *Avicennia* over an area of 208.79 ha in the AMPA zone (Photo 14). Similarly, with the collaboration

[5] Cited in the 2010 report by Adoption aux Changements Climatique côtier on strategies to combat coastal erosion.

of AZOHS, mangrove was planted by the population in the *bolongs of* Kafountine, Diana, Abéné and Katack, nearly 37,000 *Rhizophora* propagules over a total area of 47 ha (PRCM/AZOHS, 2013). Also, under the support of the NGO Justice et Développement, island populations reforested a total area of 182.18 ha with 910,949 *Rhizophora* propagules in the *bolongs of* Diogué, Niomoune, Haere, Kouba and Hilol in 2015(ONG JD, 2015).

Photo 14: Mangrove reforestation in Albadar (A) and Niomoune Island (B) (Barry, April and June 2016)

These mangrove reforestation operations, carried out by a highly aware and concerned population, have enabled the restoration of certain *bolongs* and degraded mangrove areas in the Commune, threatened by salinization and silting. However, more needs to be done on the Commune's islands, particularly Bliss, where initiatives by NGOs, government bodies and development partners are still very timid.

For sustainable mangrove management in the Commune of Kafountine, it is necessary to

- remove the misunderstanding that mangroves are an inexhaustible source of energy;
- put stilt-root cutting in damaged areas on hold to allow them to regenerate;
- encourage reforestation campaigns where ecological conditions are conducive to their development, with prior emphasis on training and advice for island populations in mangrove planting techniques.

Technical support on planting techniques, the reforestation period, the quality of seedlings and the nature of the substrate to be reforested, the reforestation stages and the monitoring of plantations remains a necessity to better understand these reforestation operations and accurately assess their impact on their current dynamics.

III. Combating salinization of rice fields

In the villages of the Commune of Kafountine, the fight against the salinization of rice fields has become a daily occurrence. In fact, rice growing is the main activity in the Diola environment and plays an important role in tradition. Today, however, this rice-growing activity is threatened by salt water, which affects soil fertility and consequently the social and cultural well-being of populations.

Deep ploughing techniques and the practice of transplanting, introduced by the local population, have enabled the establishment of some rice fields in low-lying areas. Similarly, dikes are squared off every season to retain rainwater and encourage flooding. The lowest rice fields, located above high tide level, are protected by dikes, desalinated by rainwater leaching and prolonged flooding.

Supported in 2009 by the World Food Program (WFP) as part of the fight against salinization of rice fields, the program's intervention in the Commune took place in the villages of Hilol, Saloulou, Boko, Diogué and Niomoune. Its aim is to support the local population in building the traditional dikes that surround the rice fields.

This program has produced satisfactory results in the Commune's valleys. Indeed, the dikes installed have enabled the recovery of land degraded by the effects of salt (Photo 15). According to the local population, rice-growing plots that had been abandoned for several years because of the salinity have been reclaimed.

Photo15: Traditional dike at Diogué (Barry, August 2016)

However, even if this program has saved some rice fields from salt intrusion, today most of the dikes built have collapsed due to the attack of erosion agents and lack of maintenance. This is the case in Hilol where, according to the village chief, Mr Demba, some rice fields had started to take on salt two years after the dikes were installed. He asserts that the construction of these dikes has over-salted the soil downstream from the dikes. The sand used to build the dikes and the traditional equipment (such as the "kadiandou") used in their construction do not allow for proper compaction. As a result, the dike may break or submerge under the onslaught of water and other erosive agents. In fact, the inertial forces of raindrops and the impact of runoff on steep slopes in terrace areas attack and weaken the dikes, leading to their destruction if they are not regularly and permanently maintained. As a result of this loss of arable land due to soil salinization, rice yields are falling. The corollary of this situation is the population's dependence on industrial rice, and the gradual displacement of rice growers to the uplands, sometimes with the illegal cultivation of Indian hemp, which is financially more profitable.

A return to rice-growing is therefore essential if the rice fields are to be properly managed, so as to isolate the fields and halt the advance of salinization. The installation of anti-salt dykes to reinforce existing diking systems helps to retain rainwater in the rice fields and isolate them from the sea water often brought in by high tides.

Improving these traditional structures could be the solution. However, their installation requires engineering work to be carried out by specialized companies, with a participatory approach to ensure the involvement of local populations. Monitoring mechanisms need to be put in place to ensure sustainable protection of rice-growing land.

IV. Accessibility to drinking water

Accessibility to drinking water is a major concern for the island populations of the Karone, Bliss and Petit-Kassa islands in the Commune of Kafountine. Sea water penetrates inland, affecting wells. Faced with this situation, local populations are developing local adaptation strategies to facilitate access to drinking water. In the village of Hilol, people have set up a water supply system based on their own resources. Cisterns are connected to five (5) public taps, managed by the local people and serving the entire village at all times (Photo 16).

Photo16: Hilol water supply system (Barry, April 2016)

In the island villages of Boko, Saloulou and Kailo, each family has set up a rainwater harvesting system that collects rainwater in basins and stores it in 20-liter cans. This water is used for self-consumption.

On the Karone islands, in the villages of Couba and Coumbaloulou, wells are between 6 and 11 meters deep, so access to water is less of a problem in these villages. Inhabitants of Mantate and Couba, where problems of access to water are acute, travel to Coumbaloulou to draw water in return for a fee, often paid at the end of the harvest.

In the Bliss Islands, access to drinking water is a particular problem in the village of Boko. In Boko, people sometimes travel as far as Saloulou to get water. The means of transport used to fetch water is the rowing dugout canoe.

The problem of access to drinking water is a reality in the Commune of Kafountine, but it is more acute on the islands, where saline intrusion affects the water table.

Insofar as the supply of drinking water is a major problem, the promotion of an efficient rainwater storage and conservation system becomes an asset. This system involves the installation of an impluvium or rainwater harvesting tank. This gentle, low-cost technology can be used on the islands of the Commune of Kafountine, where drinking water requirements are high. The system involves collecting rainwater from roofs and connecting it to a reservoir for storage and conservation.

Improved management of the impluviums requires the implementation of a monitoring and training mechanism for the beneficiary populations on the use and maintenance of this structure.

Conclusion

Coastal erosion in the Commune of Kafountine has increased over the last few decades. This situation has led the local population, with the support of state and non-state structures, to implement adaptation strategies. These strategies, based on structural and non-structural options, sometimes show their limits from an ecological and socio-economic point of view. To be effective, control strategies must be supervised, supported and accompanied by the various stakeholders, and the actions to be taken must meet the requirements and constraints of the environment. This will require a thorough understanding of the physical environment of the coastal zone, in order to better propose solutions and avoid costly mistakes.

GENERAL CONCLUSION

The coastal zone of the Commune of Kafountine is a geographical space in which we find a diversity of morphological landscapes and socio-economic activities that today are threatened by coastal erosion.

The use of remote sensing and cartography has enabled us, based on the diachronic analysis of Landsat satellite images from 1979, 1986, 2000 and 2015, to assess the retreat and progradation of the coastline. Generally speaking, the Abéné to Birds peninsula sector recorded an average progradation rate of (+26.96 m/year) between 1979-1986 and 1986-2000, and a recession rate of (-1.30 m/year) between 2000-2015. The sector from Ebédou to Diogué showed an accumulation trend between 1979-1986, and 1986-2000, an average progradation rate of (-14.25m/year) and an average recession rate of (-3.32 m/year) between 2000-2015.

The coastline in the Commune of Kafountine is constantly evolving, depending on its sensitivity to hydrodynamic agents (swell, longshore drift, currents, waves, etc.) and climatic factors. Human intervention has become an increasingly important factor in the evolution of this part of the coastline. The main coastal developments that have influenced the processes at play on this coast are the construction of seaside hotels, the extraction of sand for building materials in several places, and the occupation of beaches by fishing and fish processing facilities. These natural and human factors have profoundly disrupted the ecological balance of the coastal system.

The consequences of this erosion are surface losses of (202.08 ha from Abéné to the Bird Peninsula and 156.57 ha from Ebédou to Diogué), mangrove degradation of (8.51% between 2000 and 2015), loss of rice paddies of (61.37% between 1979 and 2015) and water degradation on the islands. This salinization has led to a drop in rice production and a decline in the population's income. On the islands, access to drinking water is limited and unpredictable due to saline intrusion and irregular rainfall.

Coastal erosion has also led to the destruction of tourist buildings in the coastal area of Kafountine.

As a result, adaptation strategies have been undertaken by local populations and development actors, using both traditional and modern techniques.

However, analysis of these control strategies reveals shortcomings linked to the monitoring and maintenance of structures, as well as to the low technical capacity of local players.

Today, it is important to consider the impact of coastal erosion and the sustainability of adaptation strategies in this context of permanent change. It is therefore obvious to work towards sustainable coastal management. The fight against coastal erosion must be led by specialists in coastal issues, but also by the involvement of all stakeholders in order to gain a better understanding of how the coast functions, and to avoid emergency or individual decisions that run the risk of opting for a solution whose implementation may prove a source of disastrous consequences for the future of populations.

BIBLIOGRAPHY

ABE (J.), 2005, Contribution à la connaissance de la morphologie et de la dynamique sédimentaire du littoral ivoirien (cas du Littoral d'Abidjan) : essais de modélisation en vue d'une gestion rationnelle, Thèse d'Etat Es-Sciences Naturelles, UFR des Sciences de la Terres et des Ressources Minières, Université de Cocody, 352 p

ACCC/UNESCO, 2012. Rapport du séminaire de clôture sur la Gestion Intégrée des Zones Côtières et les Changements Climatiques en Afrique de l'Ouest, Dakar/Sénégal, 5p

ADJOUSSI (P.), 2001. Impact du prélèvement du sable marin sur l'évolution du trait de côte à Yoff : ESSAI D'ETUDE DE VULNERABILITE, (Presqu'île du Cap Vert, Sénégal), Mémoire D.E.A, FLSH, UCAD, 72P.

André (P.), 1999. L'évaluation des impacts sur l'environnement, processus, acteurs et pratique, Presses internationales Polytechnique, Québec, 316 p.

AKADJE (A.M.L.M.C), 2010. Analyse de la cinématique du trait côte de Port-Oouet à l'aide du couplage télédétection et SIG (Cote d'Ivoire), DEA, Université de Cocody-Abidjan, 42 p

PRCM/AZOHS, 2013. Small initiatives for a big cause: Conserving Biodiversity, 10p

BA (M.), 1996. Rapport d'expertise sur la plage de l'hôtel Savana à Saly.Dakar, March 1996,16 p.

BIRD (E.C.F.), 1985. *Coastline changes. A global review.* John Wiley & Sons, Chichester, New York, Brisbane, Toronto, Singapore, 219 p.

BIRD (E.C. F.), 2000. Coastal geomorphology. An introduction. John WILEY and Sons Ltd, Chichester, 322 p

BOUDJERA (A.), 2010. Impact des structures hydrauliques sur l'érosion littorale en Côte- Nord. Étude de cas précis: CAS DE SEPT-ÎLES, Master's thesis in Geography, Université du Québec à Montréal, 137 p.

BLIVI (A.) 2000. Erosion côtière dans le Golfe de Guinée en Afrique de l'Ouest: Exemple du Togo, FAC des sciences humaines, département de géographie, center de gestion intégrée du Littoral et de l'environnement, 8p.

BRUNET (R.), FERRAS (R.) and THERY (H.) 2005. Les mots de la géographie: dictionnaire critique, Reclus- la documentation Française, 518 pages.

BRUUN (P.) 1962. Sea level rise as a cause of shore erosion. Journal of the Waterways and Harbors division. Proceedings of the American Society of Civil Engineers, vol. 88, pp. 117 - 130.

CHANDLER (A.), 1960. Stratégies et structures de l'entreprise, Paris, Organisation.

COMPRENDRE L'EROSION COTIERE/ ZIP, 2008. Le cahier du participant, St Siméon, 64p

CORMIER-SALEM (M-C), 1999. Les Rivières du sud, sociétés et mangroves Ouest- africaines, Volume1, IRD, Paris, 426p.

DACOSTA (H.), 1989. Precipitation and runoff in the Casamance basin. Doctoral thesis in geography. UCAD, Dakar, 283 p.

DEMOULIN (D.), 1967: Geomorphological study of the Petite Côte coastline at Bargny, UCAD, Dakar, 122p

DENNIS (K.C.), NIANG-DIOP (I.), NICHOLLS (R.J.), 1995. Sea-level rise in Senegal: Potential impacts and consequences. J.Coasral Res., Fort Lauderdale (Flo.), NICHOLLS, (R.J.), LEATHERMAN, (S.P.) (eds) "The potential impacts of accelerated sea-level rise on developing countries", Sp.Issue 14,

DIABANG (S.), 2016. Bois-énergie et fumoirs au quai de pêche de Kafountine (Basse Casamance), master's thesis, Université Assance seck de Ziguinchor, Département de Géographie, 160p

DIALLO (S.), 1982. Evolution géomorphologique du littoral sur la petite côte à Rufisque. Master's thesis, UCAD, 124 p.

DIARA (M.), 1999. Formation et évolution fini-Holocène et dynamique actuelle du Delta Saloum - Gambie (SENEGAL-WEST AFRICA), PhD thesis, University of PERPIGNAN, 161p.

DIATTA (I.), 2012. Erosion of the Casamance coastline from Nikine to Cabrousse. Master II thesis, UCAD. 98 p

DIOP (E.S.), 1975. - Etude géomorphologique de la Pointe de Sangomar et des Iles du Gandoul (Sénégal). T.E.R. - Faculté des Lettres, 183 p..

DIATTA (I.), 2004. Opening a breach across the Langue de Barbarie (Saint Louis du Sénégal). Public authorities and the consequences of the breach. Mém maitr. Univ. Gaston Berger (Sénégal), Section de Géogr. 116 p.

DIATTA (M. C. B. C.), (2007). Eau et Pauvreté dans la Communauté rurale de Kafountine, Master's thesis in geography, Université Gaston Berger de Saint-Louis, 138p.

DIATTA (A.J.) and DIOUF (D.A.), 2013. Le Mémoire : Méthodologie de recherche, normes et techniques de rédaction, conseils pour la soutenance, Imprimerie Néma, 2013, 101p.

DIAW (A.T.), 1997. "Evolution des milieux littoraux du Sénégal : Géomorphologie et Télédétection", Paris, Université de Paris I, 270 p.

DIEYE (E.H.B.), DIAW (A.T.), SANE (T.), NDOUR (N.), 2013. Dynamics of the mangrove in the Saloum estuary (Senegal) between 1972 and 2010. In Cybergeo: European Journal of Geography, Environment, 22 p.

DIEYE (E.H.B.), 2007. Les ensembles littoraux de la lagune de Joal-Fadiouth et de l'estuaire du Saloum (Sénégal): Approche méthodologique de la dynamique de la mangrove entre 1972 et 2005 par télédétection et système d'information géographique, PhD student 3e cycle, FST/UCAD, LERG, Dakar, 217 p.

DILLENBURG (S.R.), ESTEVES (L.S.), TOMAZELLI (L.J.), 2004. A critical evaluation of coastal erosion in Rio Grande do Sul, Southern Brazil. Annals of the Brazilian Academy of Science, vol. 76, no. 3, pp. 611 - 623.

NIANG-DIOP (I.), 1995. Erosion on the Petite Côte du Sénégal from the example of Rufisque. Past - Present - Future. Thesis and microfilmed documents. Published by ORSTOM. Université D'ANGERS, UER Environnement, 477p

DIOP (S.), 1990. La Côte ouest-africaine, du Saloum (Sénégal) à la Mellacorée (République de Guinée). Paris: ORSTOM, 366p.

DOMAIN, 1976. Carte sédimentologique du plateau continental sénégambien .Extension à une partie du plateau continental de mauritanien et de la guinée Bissau. ORSTOM ed ; paris, notice explicative N° 68. 17p

DREAL Haute-Normandie, 2011. Etude sur la sensibilité et sur l'adaptation de la Haute Normandie aux effets du changement climatique, Phase A report, 96p.

DUPRIEZ (H.) and DE LEENER (P H.), 1990. Les chemins de l'eau: ruissellement, irrigation, drainage (tropical manual), Terres et Vie, CTA, ENDA, l'Harmattan 380 p.

EPIPHANE (D.C.G.), 2009. Géomorphologie et érosion côtière dans le Golfe de Guinée, Master 2 thesis in Physical Oceanography, Université d'Abomey-Calavi (UAC), BENIN, Faculté des Sciences et Techniques (FAST), 100 p.

FAYE (G.), 1993. L'érosion côtière sur le littoral sud du cap vert à l'embouchure du Saloum (Sénégal). Thèse de doctorat de 3e cycle universitaire cheikh Anta Diop de Dakar, UCAD, 293p.

FAYE (I.B.N.), 2010. Coastline dynamics on sandy coastlines from Mauritania to Guinea-Bissau (West Africa): Regional and local approach by photo-interpretation, image processing and analysis of old maps. PhD thesis, Université de Bretagne Occidentale, volume 1, 321 pages

GAYE (A. T.), SYLLA (M. B.), 2008. Climate scenarios in Senegal. Laboratoire de Physique de L'Atmosphère et de l'Océan S. F. (LPAO-SF), Ecole Supérieure Polytechnique Université Cheikh Anta Diop, Dakar, Senegal.

GEORGE (P.), AND VERGER (F.) 1970. Dictionary of geography, 9th updated edition, 6, avenue Reille, 1970, 75014 Paris.

GIRARD (M.C.), GIRARD (C.M.), 2010. Traitement des données de télédétection: Environnement et ressources naturelles, 2nd edition, Dunod, Paris, 1999, 2004, 2010, 553 p.

IPCC, 2007. Climate Change 2007. Contribution of Working Groups I, II and III to the Fourth Assessment Report of the Intergovernmental Panel on Climate Change, IPCC, Geneva, Switzerland, 103p.

GOPINATH (G.), SERALATHAN(P.), 2005. Rapid erosion of the coast of Sagar island, West Bengal-India. Environmental Geology, vol. 48, n°8, p 1058 - 1067.

HAPKE (C.J.), REID (D.), RICHMOND (B.M.), RUGGIERO (P.), LIST (J.), 2006. National Assessment of shoreline change Part 3: Historical shoreline change and associated coastal land loss along sandy shorelines of the California coast. U.S. Geological Survey, Openfile Report 2006-1219, 72 p.

HOUGHTON (J.T.), JENKINS (G.J.), EPHRAUMS (J.J.), 1990. Climate change: IPCC Scientific Assessment, Cambridge, University Press.

IBE (A. C.), QUELENNEC (R. E.), 1989. "Methodology for assessment and control of coastal erosion in West and Central Africa", UNEP Regional Seas Reports and Studies, Nairobi, n° 107, 107 p.

IBE (A.C.), ANTIA (E.E.) 1983. Preliminary assessment of the impact of coastal erosion along the Nigerian Shoreline. NIOMR Tech. Paper, No. 13, 17 p

IDEE CASAMANCE, 2006. La Casamance littorale: Réserve de biosphère côtière, UNESCO/MAB, Ziguinchor, 16p.

IFEN, 2006. A quarter of the coastline is retreating due to erosion. IFEN 4 pages, n°13, 4 p

KALCK (Y.), 1978. Evolution of mangrove areas in Senegal in the recent quaternary. Geological and geochemical studies. Thèse 3e cycle, ULP Strasbourg, 117 p.

KLEIN (J), 2003. Protecting the coastline in French overseas departments. PhD thesis, ENS LSH-Laboratoire "espace et culture", Paris IV, Sorbonne, 257 p.

LACOSTE (Y.), 2003. De la géopolitique aux paysages: dictionnaire de la géographie, Armand Colin, 413 p.

LOYER (J.Y.), BOIVIN (P.), LE BRUSQ (J.Y.), ZANTE (P.), 1986. Les sols du domaine fluviomarin de Casamance (Sénégal): Evolution récente et réévaluation des contraintes majeures pour leur mise en valeur, ORSTOM, Dakar-Sénégal, in: Selected Papers of the Dakar Symposium on Acid Sulphate Soils, Dakar, Senegal, January, 1986, Edited by H. Dost, pp. 16-23.

LUSSEAULT, 2001. Analyse stratégique, Ellipses, p. 86.

LY (C.K.), 1980. The role of the Akossombo dam on the Volta River in Causing Coastal erosion in central an eastern Ghana. Marine Geology, 37 (3/4), pp. 323.332.

MAHÉ (G.), 2006. Variabilité pluie-débit en Afrique de l'ouest et centrale au 20ème siècle : Changements hydroclimatiques, occupation du sol et modélisation hydraulique. Université des sciences et techniques, Montpellier 2, 160p.

MATE AND METL ,1997. Plan de prévention des risques littoraux. La documentation française, Paris, 54 p

MARIUS (C.) ,1986. Evolution of the Gulf of Casamance in the recent Quaternary and changes in mangrove vegetation and soils associated with the present drought. ORSTOM, 296p

MCCARTHY (J.J.), CANZIANI (O.F.), LEARY (N.A.), DOKKEN (D.J.), WHITE (K.S.), 2001. Climate change 2001: Impacts, adaptation and vulnerability. Cambridge University. Press, Cambridge, 1005p.

MONTOROI (P), 1991. Gestion d'un barrage anti-sel en Basse Casamance (Sénégal) Département Eaux Continentales de I'ORSTOM, ORSTOM, BP 1386, DAKAR, Sénégal,. p275-285

MORTON (R.A.), PAINE (J.G), (1985). Beach and vegetation-line changes at Galveston Island, Texas: Erosion, deposition, and recovery from Hurricane Alicia. Geological circular, vol. 85-5, 39 p.

MORTON (R.A.), SPEED (M.F), (1998). Evaluation of shorelines and legal boundaries controlled by water levels on sandy beaches. Journal of Coastal Research, vol. 14, n°4, p. 1373 - 1384.

MORTON (R.A.), MILLER (T.L.), MOORE (L.J.), 2004. National assessment of shoreline change: Part 1. Historical shoreline changes and associated land loss along the U.S. Gulf of Mexico. U.S. Geological Survey, Open-file report 2004-1043, 42 p.

MORTON (R.A.), MILLER (T.L), 2005. National assessment of shoreline change: Part 2. Historical shoreline changes and associated coastal land loss along the U.S. Southeast Atlantic Coast. U.S. Geological Survey Openfile Report 2005-1401, 35 p.

MOUGENOT (B.), 1996. Etude par Télédétection des états de surface en relation avec les nappes, exemples des sols salés et acidifiés de Basse Casamance (Sénégal), ORSTOM, 8p.

NDAO (M.), 2012. Dynamiques et gestion environnementales de 1970 à 2010 des zones humides au Sénégal: étude de l'occupation du sol par Télédétection des Niayes avec Djiddah Thiaroye Kao (à Dakar), Mboro (à Thiès) et Saint-Louis, PhD thesis, University of Toulouse, 371 p.

NIAZI. (S.), 2007. Evaluation of the impacts of climate change and sea-level rise on the Tetouan coastline (Western Mediterranean of Morocco): Vulnerability and Adaptation. PhD thesis, Faculty of Sciences, Mohamed V- AGDAL University, 296 p.

NICHOLS (R.J.), 1998. Assensing erosion of sandy beaches due to sea-level rise. In: Geohahazard in Engineering Geology (Maund, J.G. and M. Eddleston (eds)). Engineering Geology special

Publications, Geology Society, London, United Kingdom, 15, pp. 71-76

OECD, 1993. Coastal Zone Management. Politiques intégrées, OECD Publishing, Paris, 144 p.

ONG JD, 2015. Rapport Final de reboisement de mangrove et de plantation de filaos en Basse Casamance dans les îles de Carabane, Diogué, Niomoune, Haere, Kouba et Hilol, 38 p

UNIDO, 1998. Profil environnemental du littoral du Togo. 71 p.

PASKOFF (R.), (1983). Beach erosion. La Recherche, Paris, 14 (140), pp. 20-28,

PASKOFF (R.), 2001. Sea-level rise and coastal areas. Myth and reality. Institut Océanographique, collection "Propos", 190p.

PELISSIER (P.), 1966. Les Paysans du Sénégal. Les civilisations agraires du Cayor à la Casamance. Thèse de Doctorat d'Etat, Paris, 939p.

PENNOBER (G), 1999. Analyse spatiale de l'environnement Côtier de l'Archipel des Bijagos (Guinée Bissau), PhD thesis, Université de Bretagne Occidentale, 232 p.

PIGNAL (J.Y.), 2005. Atelier Changement Climatique : les petites îles face au changement climatique, Ecole Nationale des Ponts et Chaussées, 38p.

PIRAZOLLI (P A.), 1993. Les littoraux. Impact des aménagements sur leur évolution. Coll. "Géographie d'aujourd'hui". Paris, pp 37-49.

PNUE/UNESCO/ONU-DAESI, 1985. Coastal erosion in West and Central Africa. UNEP Regional Seas Reports and Studies n°67, 248 p.

REBERT (J.P.), 1983. Hydrology and water dynamics of the Senegalese continental shelf.
Doc. Scient, ISRA/CRODT, DAKAR, 89-99p.

SALL (M.M.), 1971. Dynamique et morphogenèse actuelle (contribution à l'étude géomorphologique du Sénégal occidental). 3rd cycle thesis. Faculté des Lettres et Sciences Humaines, Department of Geography, 290p.

SALL (M.), 1982. Dynamique et morphogenèse actuelles du Sénégal occidental, Thèse de Doctorat d'Etat ès-Lettres, 2 volumes, Université Louis Pasteur, Strasbourg, 604p.

SAVADOGO (S.), 2013. Conséquence de l'érosion côtière sur le littoral ivoirien, Brevet de Technicien Supérieur, Ecole Supérieur de Technologie LOKO, 42p.

SARR (S.), 2012. Insular vulnerability in the Southern Rivers (Atlantic Casamance). Master's thesis, UGB, 135p.

SECK. (O.), 2001. Les stratégies de lutte contre l'érosion dans la Commune rurale de Diass, Geography thesis, UCAD, 124p.

SOUMARE (A.), 1992. Evolution géomorphologique récente des paysages du Bas-Saloum. DEA dissertation. Dakar: Institut des Sciences de l'Environnement, Université Cheikh Anta Diop, 61p.

SOUMARE (A.), 1996. Etude comparative de l'évolution géomorphologique des bas-estuaires du Sénégal et du Saloum (approche par les données de terrain et de télédétection). Doctoral thesis, 3rd cycle in geography, UCAD, 265p.

SY (B.A.), SY (A.A.), 2010. Dynamique actuelle du cordon littoral de la Grande Côte sénégalaise de Saint Louis à Niayam et ses conséquences ; RGLL, N°08, 18p.

SY (A.A.), 2013. Dynamiques sédimentaires et risques actuels dans l'axe Saint-Louis-Gandiol, littoral Nord du Sénégal, PhD thesis, Université Gaston Berger de Saint-Louis, 328p.

TALL (E.S.B.), 2013. La dynamique de la mangrove et ses impacts environnementaux et socioécomiques à Diogué en Basse Casamance de 1979 à 2010, master's thesis, Université Assance seck de Ziguinchor, Département de Géographie, 98p.

THIAM (M.T.), CROWLEY (J.), 2014. Impact of environmental change on human migration: Case study: Senegal and Cote d'Ivoire, UNESCO, 294p.

THIOR (M.), 2013. Impacts environnementaux et socioéconomiques de l'érosion côtière dans la Commune Rurale de Diembering (Basse Casamance), Master's thesis, Université Assane Seck, 132p.

European Union, 2004. Living with coastal erosion in Europe: Space and sediment for sustainable development. Assessment and recommendations for sustainable development. Assessment and recommendations of the Erosion project.

VIEILLEFON (J.), 1975. Carte pédologique de la Basse Casamance (domaine fluvio-marin) à 1/100 000, notice explicative No 57, O.R.S.T.O.M, 67p.

WEBOGRAPHY

http://www.earthexplorer.gov.

http://www.ideecasamance.org/casamancefr.htm http://www. documentation.ird.fr

http://www.unesco.org

http://www.memoireonline.org

http://www.ramsar.org

http://www.prcmarine.org

APPENDIX1: QUESTIONNAIRE

Erosion côtière et impacts dans la Commune de Kafountine

2015/2016 - Etudiant
Mars 2016

I.IDENTIFICATION

1. Nom et Prénom

2. Nom du village

3. Quel est votre âge?

4. Niveau d'instruction
☐ 1. Analphabète ☐ 2. Coranique ☐ 3. Primaire
☐ 4. Moyen ☐ 5. Secondaire ☐ 6. Supérieur
Vous pouvez cocher plusieurs cases (2 au maximum).

5. Quelle est votre profession?
☐ 1. Paysan ☐ 2. Fonctionnaire ☐ 3. Pêcheur
☐ 4. Ouvrier ☐ 5. Autre
Vous pouvez cocher plusieurs cases (2 au maximum).

6. Quelle est le type de construction de votre maison?
☐ 1. Traditionnel ☐ 2. Moderne ☐ 3. Mixte
Vous pouvez cocher plusieurs cases (2 au maximum).

II. FACTEURS DE L'EROSION COTIERE

7. Avez-constatez l'érosion côtière?
☐ 1. Oui ☐ 2. Non
Vous pouvez cocher plusieurs cases.

8. Si oui, A quelle période de l'année?

9. Quels sont d'après vous les zones de votre localité les plus touchés par ce phénomène?

10. Existe-t-il des indicateurs de ce phénomène?

11. Selon vous qu'est ce qui est à l'origine de ce phénomène?

12. Existe-t-il une relation entre vos activités quotidiennes et l'apparition de ce phénomène?
☐ 1. Oui ☐ 2. Non
Vous pouvez cocher plusieurs cases.

13. Si oui, comment?

II. IMPACTS ENVIRONNEMENTAUX DE L'EROSION COTIERE

14. Quels sont les impacts de l'érosion côtière sur la végétation?

15. A quel rythme pouvez-vous qualifier cette situation?
☐ 1. Lent ☐ 2. Moyen ☐ 3. Rapide
Vous pouvez cocher plusieurs cases (2 au maximum).

16. Quelles sont les espèces les plus menacées?

17. Avez-vous eu connaissance ou senti la présence de la salinisation dans votre milieu ?
☐ 1. Oui ☐ 2. Non
Vous pouvez cocher plusieurs cases.

18. Si Oui ; comment pouvez- vous identifier le phénomène?
☐ 1. Réduction du couvert végétal
☐ 2. Appauvrissement des sols
☐ 3. Apparition de couches blanchâtres
☐ 4. Autres
Vous pouvez cocher plusieurs cases (2 au maximum).

19. " Autres " à préciser

20. La salinisation dans votre zone est du à quoi?
☐ 1. Déficit pluviométrique
☐ 2. Evaporation
☐ 3. Intrusion marine
☐ 4. Surexploitation des terres
☐ 5. Déboisement
☐ 6. Exploitation du sel
Vous pouvez cocher plusieurs cases (2 au maximum).

21. Comment pouvez-vous apprécier l'ampleur du phénomène dans votre zone?
☐ 1. Lent ☐ 2. Moyen ☐ 3. Rapide
Vous pouvez cocher plusieurs cases (2 au maximum).

3.1. Impact sur la riziculture

22. Avez- vous des terres rizicoles?
☐ 1. Oui ☐ 2. Non
Vous pouvez cocher plusieurs cases.

23. Si oui, où se situent-elles ?

24. Vos terres sont-elles menacées par la salinisation?
O 1. Oui O 2. Non

25. Avez-vous perdus ou abandonné des champs à cause de la salinisation ?
O 1. Oui O 2. Non

26. Si oui, combien de vos terres avez-vous perdu?
O 1. 1 O 2. 2 O 3. 3 O 4. 4
O 5. 5 O 6. plus de 6

27. Combien sont toujours cultivables?
O 1. 1 O 2. 2 O 3. 3 O 4. 4
O 5. 5 O 6. plus de 6

28. Cette situation n'entraine-t-elle pas la baisse de la production rizicole ?
O 1. Oui O 2. Non

29. En quelle quantité ?
O 1. Le tiers O 2. La moitié O 3. plus que la moitié

30. Comment comptez-vous faire face à cette situation?

3.2. Impact sur la qualité de l'eau

31. Quelle eau consommez-vous?
☐ 1. robinet ☐ 2. puit ☐ 3. autre
Vous pouvez cocher plusieurs cases (2 au maximum).

32. L'avancée de la mer n'affecte-elle pas la qualité de l'eau des puits?
☐ 1. Oui ☐ 2. Non
Vous pouvez cocher plusieurs cases.

33. Si oui, comment l'expliquez-vous?

34. Quel est le goût de l'eau consommée?
O 1. Salé O 2. Saumâtre O 3. Autre

35. Avez-vous abandonné des puits pour cause de la salinité?
☐ 1. Oui ☐ 2. Non
Vous pouvez cocher plusieurs cases.

36. Quelle est la qualité de l'eau des puits?
O 1. Très bonne O 2. Bonne O 3. Mauvaise
O 4. Très mauvaise

37. Comment comptez-vous faire face à cette situation ?

3.3. Risque de submersion

38. Etes-vous confrontés à des problèmes d'inondation ?
O 1. Oui O 2. Non

39. Si oui, depuis quand?

40. Connaissez-vous des maisons et/ou des infrastructures menacées de submersion ?
☐ 1. Oui ☐ 2. Non
Vous pouvez cocher plusieurs cases.

41. Si oui, lesquels ?

42. Avez-vous pensé à se déplacer en cas de forte avancée de la mer ?
☐ 1. Oui ☐ 2. Non
Vous pouvez cocher plusieurs cases.

3.4. Impact sur la pêche

43. Est ce que les activités de la pêche sont affectés par l'erosion cotiere ?
☐ 1. Oui ☐ 2. Non
Vous pouvez cocher plusieurs cases.

44. Si oui, quelles sont les installations les plus affectées?

45. Y'a-t-il une possibilité de déplacer ces installations?
☐ 1. Oui ☐ 2. Non
Vous pouvez cocher plusieurs cases.

46. Si Non pourquoi ?

3.5. Impact sur le tourisme

47. Est-ce que les aménagements touristiques sont affectés par l'avancée de la mer?
☐ 1. Oui ☐ 2. Non
Vous pouvez cocher plusieurs cases.

48. Si oui, quelles sont hôtels et les campements les plus affectés?

49. Y'a-t-il une possibilité de déplacer les hôtels et campements?
☐ 1. Oui ☐ 2. Non
Vous pouvez cocher plusieurs cases.

50. Si non pourquoi?

51. Cette situation n'est-elle pas une menace sur les activités touristiques ?
☐ 1. Oui ☐ 2. Non
Vous pouvez cocher plusieurs cases.

52. Si oui, comment ?

IV. STRATEGIES D'ADAPTATION DE L'EROSION COTIERE

53. Avez-vous entrepris des actions d'adaptation?
☐ 1. Oui ☐ 2. Non
Vous pouvez cocher plusieurs cases.

54. Si oui, lesquelles?

55. Quels sont les acteurs?

56. Quelles sont les techniques mises en place pour lutter contre l'érosion côtière?

57. Si sont ceux du reboisement quelles sont les espèces qui ont été reboisé ?

58. Depuis quand?

59. Ont-elles survécu?
○ 1. Oui ○ 2. Non

60. Si " non " pourquoi ?

61. Quelles appréciations faites- vous des réalisations?
☐ 1. Très efficaces ☐ 2. Passables ☐ 3. Inefficaces
Vous pouvez cocher plusieurs cases (2 au maximum).

62. Quelles sont les difficultés rencontrées dans les stratégies?

63. Existe-t-il un suivi mis en place ?
☐ 1. Oui ☐ 2. Non
Vous pouvez cocher plusieurs cases.

64. Si " Oui " lesquelles?

65. Quelles perspectives préconisez-vous?

APPENDIX 2: Interview guide

INTERVIEW GUIDE FOR THE HEAD OF THE WATER AND SEWAGE BRIGADE DIOULOULOU FORESTS AND THE AMPA CONSERVATORY

I. **Manifestations of coastal erosion on coastal vegetation**

How do you explain the decline in coastal vegetation?

What are the consequences for biodiversity?

II. **Coastal erosion control strategies**

What are your strategies for combating the loss of coastal plant cover?

Do you benefit from government support for the preservation of coastal plant cover?

Are there any partners involved in protecting coastal vegetation?

How do you rate these interventions?

MAINTENANCE GUIDES FOR HOTELIERS

I. Impact of coastal erosion on tourism

How does the sea affect your beach?

Are your hotels or campsites threatened by the advancing sea? If so, what are the impacts?

II. Coastal erosion control strategies

Have you undertaken any strategies to combat the phenomenon?

If so, what are these actions?

INTERVIEW GUIDE FOR THE MAYOR OF KAFFOUNTINE

I. Impacts of coastal erosion

How do you feel about the advancing sea?

What do you think are the causes?

What are the most remarkable consequences?

II. Coastal erosion control strategies

Have you developed initiatives to combat the phenomenon?

If so, are they local or state-run?

What are these actions?

Are development partners involved in control strategies?

What do you think of these actions?

What do you recommend for more effective coastal erosion control strategies?

MIX
Papier aus verantwortungsvollen Quellen
Paper from responsible sources
FSC® C105338

Printed by Books on Demand GmbH, Norderstedt / Germany